TEST BANK

CALCULUS

from Graphical, Numerical, and Symbolic Points of View

Arnold Ostebee and Paul Zorn

Scott Inch

Bloomsburg University of Pennsylvania

Saunders College Publishing

Harcourt Brace College Publishers

Fort Worth   Philadelphia   San Diego   New York   Orlando   Austin
San Antonio   Toronto   Montreal   London   Sydney   Tokyo

Inch: Test Bank to accompany *Calculus from Graphical, Numerical and Symbolic Points of View, First Edition, Vol. 1 & 2*. Ostebee.

ISBN 0-03-017418-X

567 021 987654321

Preface

This Test Bank is a supplement to accompany *Calculus from Graphical, Numerical, and Symbolic Points of View* by Arnold Ostebee and Paul Zorn. It contains over 450 questions arranged by section and chapter. The code in the margin next to each question indicates the level of difficulty (1 is easiest, 3 is hardest) and point of view (G is graphical, N is numerical, and S is symbolic). Questions that require technology state so explicitly. Answers are found at the end of the manual.

This Test Bank is available in computerized form in IBM® PC (3 1/2"), Macintosh®, and Windows® formats. The EXAMaster+™ Computerized Test Bank allows you to preview, select, edit, and add items to tailor tests to your course or select items at random; add or edit graphics (MS-DOS version), and print up to 99 different versions of the same test and answer sheet. The On-Line Testing feature allows you to create tests and administer them over a network. These tests can then be scored and the grades recorded using the ESAGrade® gradebook software (available with the IBM and Windows versions). If you are interested in receiving a copy of the computerized test bank or would like more details about its features, please contact your Saunders sales representative.

If you have any comments, suggestions, or corrections for this test bank, please address your correspondence to: Mathematics Editor, Saunders College Publishing, The Public Ledger Building, Suite 1250, 150 South Independence Mall West, Philadelphia, PA 19106.

Acknowledgments

Many of the test questions in this manual are based on or adapted from tests given by current and former class-testers of Ostebee/Zorn's text. Special thanks are due to:

Janet Andersen (Hope College), Don Bean (College of Wooster), Joyce Becker (Luther College), Paul Bialek (College of Wooster), Manfred Boos (Concordia University), Fred Brauer (University of Wisconsin), David Bressoud (Macalester College), Carolyn Chapel (Luther College), Ralph Czerwinski (Millikin University), Ted Erickson (Wheeling Jesuit College), Joe Fiedler (California State University - Bakersfield) Mary Anne Gerlach (University of Wisconsin - Whitewater), Bonnie Gold (Wabash College), Carol Harrison (Susquehanna University), Kevin Hastings (Knox College), Phoebe Judson (Trinity University), Peter Kelly (St. Alban's School), Gary Klatt (University of Wisconsin - Whitewater), Nicholas Kuhn (University of Virginia), Steve Kuhn (University of Tennessee - Chattanooga), Tom Linton (Western Oregon State College), Norm Loomer (Ripon College), Neal Madras (York University), Susan McLoughlin (Union County College), Richard Mercer (Wright State University), Ed Packel (Lake Forest College), Ed Rozema (University of Tennessee - Chattanooga), Marsha Schoonover (Chattanooga State Technical Community College), Rod Smart (University of Wisconsin), Jim Vance (Mercer University), Joe Van Wie (Southwestern State University), Jim Young (Clackamas High School)

The test questions and their answers were carefully checked for accuracy by Susan Fettes (SUNY - Oswego) and Art Richert (Southern Adventist University). Their attention to detail and helpful suggestions are greatly appreciated. Any remaining errors are my own; corrections should be sent to the publisher at the address above.

Scott Inch
Bloomsburg University of Pennsylvania

November 1996

Dedication

To my parents,

Barbara and Elwood

CHAPTER 1: FUNCTIONS IN CALCULUS

SECTION 1: FUNCTIONS, CALCULUS-STYLE

Problems 1 and 2 refer to the following graph of a function f:

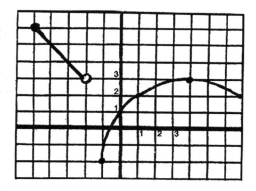

S-2   1. Find an algebraic formula for f. The graph consists of a straight line segment and a portion of a circle.

G-1   2. Find $f(-2)$, $f(0)$, and $f(4)$.

G,S-2 3. Let f be the function defined by $f(x) = x^2$. Using f, define the function g by:

$g(x)$ = the slope of the straight line through $\left(x, f(x)\right)$ and $\left(x+1, f(x+1)\right)$.

For instance, $g(1)$ = the slope of the line from (1, 1) to (2, 4) = 3.

a) Find $g(-1)$, $g(0)$, $g(1/2)$, $g(3)$.

b) Write a formula for $g(x)$ in terms of x.

c) Draw a graph of g on the interval $[-3, 3]$.

N,S-1 4. Let $f(x) = x^2$. Define the function $A(x)$ by:

$A(x)$ = the average of the value of $f(3)$ and the value of $f(x)$.

a) Find $A(0)$

b) Find $A(5)$

c) Find $A(3)$

d) Find $A(-1)$

e) Find a formula for $A(x)$

SECTION 2: GRAPHS

S-1   5. Suppose $g(x) = \log_7(x-5) + 1$. Describe the graph of g in terms of the graph of $f(x) = \log_7 x$.

G-1   6. The graph of f is shown below. Use it to answer the following problems: (Note: Answers may vary.)

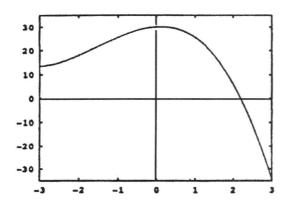

a) Find a number L such that $L < f(x)$ for x in the interval $[1, 2]$.

b) Find a number U such that $f(x) < U$ for x in the interval $[-2, 0]$.

c) Find an interval over which $-10 \le f(x) \le 10$.

d) Find an interval over which $|f(x) + 10| \le 10$.

G-1   7. Explain how to obtain the graph of $y = 5 + \ln(3x)$ from the graph of $y = \ln(x)$ using *only* vertical translation.

G-1   8. a) Sketch a graph which continues to increase at an increasing rate.

b) Sketch a graph which continues to increase at a decreasing rate.

G-1   9. The graphs below describe the learning curves of two different types of jobs. If t is the time on the job, $f(t)$ describes the fraction of the required job skills the individual possesses. One type of job is skilled or semi-skilled, such as a computer programmer. The other type of job is primarily unskilled, such as the french fry chef at a fast food restaurant. Which of the graphs describes which learning curve, and why?

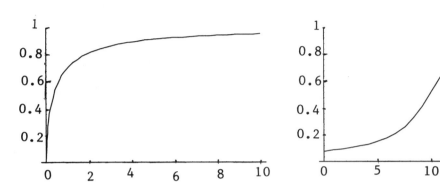

G-1   10. Let g be a function with domain all real numbers. Further, suppose g has the values shown in the table below:

x	-5	-3	-2	1	4
$g(x)$	-1	0	-4	2	1

Sketch a graph of g on the interval $[-5, 4]$. There is no unique sketch.

G-1   11. Consider the following statement: "By 1990, the cost of living in Galesburg was still increasing, but at a decreasing rate." Which of the following curves is most likely to represent the graph of cost of living as a function of time? Explain your answer.

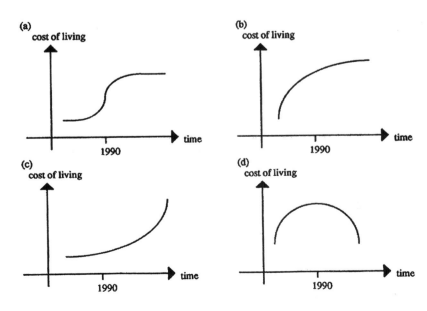

3

SECTION 3: MACHINE GRAPHICS

Using technology, graph $g(x) = x \cos x$ on $[-2\pi, 2\pi]$. Answer questions 12 and 13 based on your graph.

G-1   12. Estimate (by zooming) one of the local maximum values (accurate to within $\pm .1$). Describe your final viewing window.

G-2   13. Find an interval on which $p(x) = x$ is a good approximation to g (accurate to within $\pm .01$). (Note: Answers may vary.)

G-1   14. Use technology to approximate the roots (accurate to within $\pm .1$) of
$$f(x) = x^3 - x - 11.$$

G-1   15. Use technology to estimate the roots (accurate to within $\pm .1$) of
$$f(x) = 2x^3 - 4x + 5.$$

G-2   16. Using technology, graph the function $g(x) = \dfrac{x}{x^2 + 1}$ for x in [0, 3].

    a) Does the graph have any local maxima or local minima?

    b) If you answered "yes" to part (a), use technology to approximate the location and value of any local maxima or local minima (accurate to within $\pm .1$).

SECTION 4: WHAT *IS* A FUNCTION?

Using technology, graph $g(x) = x \cos x$ on $[-2\pi, 2\pi]$. Answer questions 17 and 18 based on your graph.

G-1   17. Is g even or odd? Explain.

G-1   18. Is g periodic? Explain.

G-1   19. Use technology to find upper and lower bounds on $f(x) = x^2 - 5$ on the interval $[-3, 2]$.

G-2   20. Using technology, find the period of the function $f(x) = \sec 2x$.

G-1   21. Find the domain and range of the function f, whose graph is given below:

4

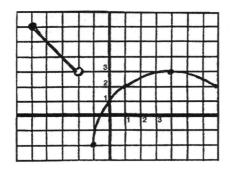

S-2   22. Let f be an even function and g be an odd function. Show that (fg) is odd. Note: $(fg)(x) = f(x) \cdot g(x)$.

G-1   23. Using technology, graph $f(x) = \dfrac{x + \sin x}{\cos x}$, for $-\pi/2 < x < \pi/2$. Is f an even function, an odd function, or neither? Explain.

S-3   24. Suppose f is a periodic function with period $1/4$ and with $f(3) = 4$, $f(3/8) = 2$. Evaluate:

    a) $f(2)$

    b) $f(11/8)$

    c) $f(x+2) - f(x)$

    d) Is it true that $f(x - 1/4) + f(x) = 2f(x)$? Explain your answer.

G,S-2   25. Is $y = 2x^6 - 4x^4 + |3x|$ even, odd, or neither? Give a reason for your answer.

G,S-2   26. Is $y = 2|x| + 6$ even, odd, or neither? Give a reason for your answer.

SECTION 5: A FIELD GUIDE TO ELEMENTARY FUNCTIONS

G-2   27. Tell whether $\sin x - \cos x$ is even, odd, or neither. Give evidence to support your answer.

G-2   28. Using technology, graph the function $2\cos(3x - 3)$. Tell what the period is by looking at the graph.

S-2   29. Find an algebraic expression for a rational function that has vertical asymptotes at $x = -2$ and $x = 5$, and has the line $y = 3/4$ as a horizontal asymptote.

G-2   30. Consider the function defined by the table of values below.

x	-3	-2	-1	0	1	2	3
$f(x)$	0	-1	0	-1	0	-1	0

a) Is the function even, odd, or neither? Explain your answer.

b) Is the function symmetric about the y-axis? Explain your answer.

c) Is the function symmetric about the origin? Explain your answer.

SECTION 6: NEW FUNCTIONS FROM OLD

S-1   31. Let $h(x) = x^3 - 4$. Find a rule for the inverse function $h^{-1}(x)$.

S-1   32. Let $h(x) = \left(\sin(2x + 1)\right)^3$.

a) Write h as a composition of two "simpler" functions.

b) Write h as a composition of three "simpler" functions.

S-2   33. Suppose that f is an odd function, and g is an even function. Let $h(x) = (f \circ g)(x)$. Is h even, odd, or neither? Explain.

S-2   34. If $f(x) = \ln x$, $g(x) = x^2 + 2$, and $h(x) = x + 2$, write a formula for:

a) $f \circ g \circ h$

b) $g \circ f \circ h$

G-1   35. Using technology, graph $f(x) = \dfrac{x + \sin x}{\cos x}$, for $-\pi/2 < x < \pi/2$. Does $f(x)$ have an inverse (in other words, does $f^{-1}(x)$ exist)? Explain.

S-3   36. Let $f(x) = \cos(x)$ and $g(x) = x^3$. Write each of the following functions as a composition of the functions f and g.

    a) $\cos(x^3)$

    b) x^9

    c) $\cos^3(x^3)$

    d) $\cos^9(x)$

SECTION 7: MODELING WITH ELEMENTARY FUNCTIONS

G-1   37. Suppose average price of a new car is modeled by $f(t) = 1350\,e^{.07606t}$, where t is the number of years after 1960 (Thus, $f(0)$ is the average car price in 1960.). Use technology to graph f and tell in what year the average new car will cost \$50,000.

N-3   38. Suppose average price of a new car is modeled by $f(t) = 1350\,e^{.07606t}$, where t is the number of years after 1960. (Thus, $f(0)$ is the average car price in 1960.) Calculate the ratio $\dfrac{f(t+10)}{f(t)}$ for the values of t given in parts (a) - (c).

    a) $t = 0$

    b) $t = 25$

    c) $t = 36$

    d) Describe each ratio calculated above in terms of a comparison between average car prices in two different years.

    e) What do you notice about the ratios?

    f) What does this tell you about the average car price?

N-2   39. An extensive study of phone usage on weekends in Dillon City shows that the function $f(t) = 1500\sin(\frac{\pi}{12}t + 9) + 1600$, where t represents the number of hours after midnight on Friday night, provides a reasonable model. Using technology, estimate:

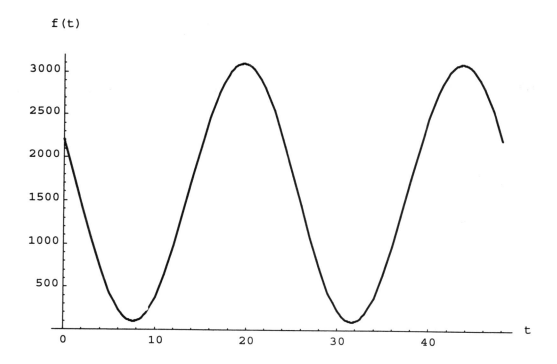

a) the number of calls made at 4:30 p.m. Sunday night

b) the time on Saturday night when the most phone calls are made

c) the length of time between successive peaks in the curve

CHAPTER 2: THE DERIVATIVE

SECTION 1: AMOUNT FUNCTIONS AND RATE FUNCTIONS: THE IDEA OF THE DERIVATIVE

S-1   1. A bullet is shot vertically from the ground at an initial velocity of 2240 ft / sec.

    a) How high is the bullet after 10 seconds?

    b) What is the velocity of the bullet after 10 seconds?

    c) What is the speed of the bullet after 100 seconds?

    d) What is the maximum height the bullet attains?

    e) How many seconds before the bullet returns and strikes the ground?

G-2   2. The function $f(x)$ is an even function. The tangent to the graph of f at $x = 4$ has slope $= -7$ and crosses the x-axis at $x = 10$. Find:

    a) $f(4)$

    b) $f'(4)$

    c) $f(-4)$

    d) $f'(-4)$

N-2   3. At 2:00 pm a car pulls onto I-75 heading north from Chattanooga. Let

$$D(t) = \text{the car's distance, in miles, north of Chattanooga, and}$$

$$V(t) = \text{the car's northward velocity, in miles per hour,}$$

where t is the number of hours after 2:00 pm.

    a) If $D(1) = 50$ and $V(1) = -60$, describe where the car is, its direction and speed at 3:00.

    b) Would it be possible for $D(2)$ to be 100 and $V(2)$ to be 20? Explain why or why not.

c) If the car makes a U - turn at 4:00, find $V(2)$.

d) Explain the relationship between $V(t)$ and $D(t)$. What does $V(t)$ tell us about $D(t)$?

e) Explain what the car is doing when $V'(t) = 0$.

N-1   4. Use Newton's Laws of Motion to answer the following questions:

$$h(t) = h_0 + v_0 t - 16t^2, \quad v(t) = v_0 - 32t$$

A ball is thrown upward from a height of 100 ft. with an initial velocity of 48 ft. per second.

a) When will the maximum height be reached?

b) What is the maximum height?

N-2   5. The function $G(s)$ gives the number of miles per gallon, G, for a 1993 Saturn SL2 as a function of its speed, s.

a) Interpret the statement $G'(55) = -0.23$ in terms of gas mileage and speed. Explain your answer.

b) If $G(55) = 30.7$, estimate $G(60)$. Explain your method.

N-1   6. Suppose $f(0) = 12$ and $|f'(x)| \le 1$. What are the possible values of $f(5)$? Justify your answer.

N-2   7. A function f has the properties:

$1 \le f'(x) \le 4$ for all x and $f(0) = 0$.

a) Explain why $2 \le f(2) \le 8$ and $5 \le f(5) \le 20$.

b) Is it possible to have $f(2) = 6$ and $f(5) = 8$? Justify your answer by either displaying (analytically or graphically) such a function or citing a theorem which shows it is not possible.

N-3   8. Suppose that f is a function with the following properties:

$f''(x) \ge -3$ for every x; $f(0) = 0$; and $f'(0) = 7$.

Show that $f(2) \geq 2$. (Optional Hint: Apply the racetrack principle twice: first apply it to f', conclude that $f'(x) \geq 1$ for every x in [0, 2], and then apply it to f.)

N-2   9. The graph of f' is shown below. Use the Racetrack Principle to find a lower bound for $f(3) - f(-1)$.

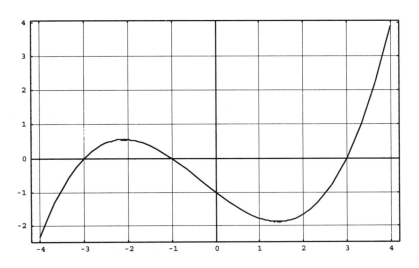

SECTION 2: ESTIMATING DERIVATIVES: A CLOSER LOOK

G-2   10. Using technology, graph $g(x) = x \cos x$ on $[-2\pi, 2\pi]$. Estimate (by zooming) $g'(\pi/2)$ to two decimal places. Describe what the graph looks like in your final viewing window.

G-1   11. Estimate $f'(6)$ using the graph of f given below:

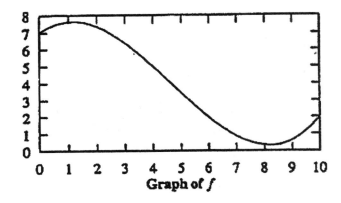

Graph of f

G-3   12. The graph of a function f appears below.

11

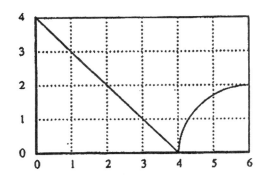

a) Using the graph, estimate $f'(5)$.

b) Using the graph, estimate $f'(2)$.

c) Using the graph, estimate $f'(4)$. Explain your answer.

d) Sketch the graph of $f'(x)$.

G-2   13. The graph of a function f is shown below:

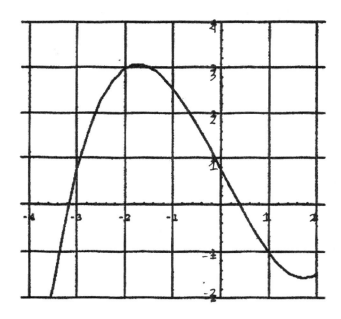

a) Estimate the missing entries in the table below.

x	-3	-2	-1	0	1	2
$f(x)$	0.75	3		0.75	-1	-1.5
$f'(x)$		0.67	-1.33		-1.33	0.67

b) Draw a tangent line at $x = -3$. What rise and run did you use to estimate the slope?

c) Draw a tangent line at $x = 0$. What rise and run did you use to estimate the slope?

14. Let f be the function whose graph is given below.

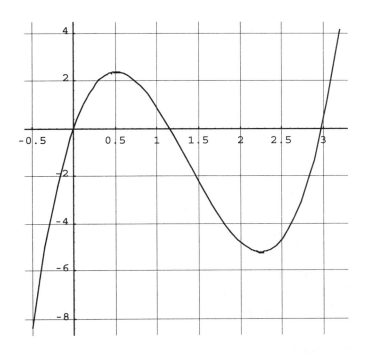

a) Fill in the following table by estimating values using the graph of f given above. Note: Answers may vary.

x	-0.5	0	0.5	1.0	1.5	2.0	2.5	3.0
$f(x)$	-8.4		2.4	0.9		-4.8	-4.7	
$f'(x)$	24.2	10		-5.3	-6.4		4.4	16.3

b) Draw the tangent line to the curve at $x = 2$.

c) Find the equation of the tangent line to the curve at $x = 2$.

13

SECTION 3: THE GEOMETRY OF DERIVATIVES

G-2   15. The graph of a function f on $[-0.5, 3.2]$ is given below. Note: f has a local maximum at $x \approx .51$, a local minimum at $x \approx 2.25$, and an inflection point at $x \approx 1.38$.

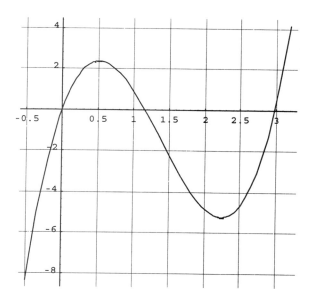

a) On what intervals is f' positive?

b) On what intervals is f' negative?

c) On what intervals is f' increasing?

d) On what intervals is f' decreasing?

e) Where does f' achieve a minimum?

f) Approximate f' at this value.

g) Approximate $f'(0)$.

h) Approximate $f'(1.5)$.

G-3   16. The graph of f' appears below. Note: f' has roots at $x \approx -0.56$, $x \approx 0.36$, and $x = 1.0$; f' has stationary points at $x \approx -0.19$ and $x \approx 0.72$; f' has an inflection point at $x \approx 0.27$.

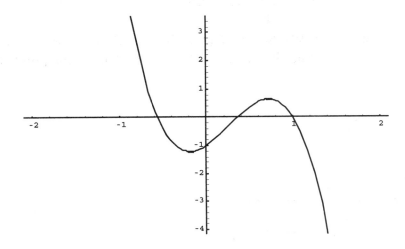

a) On what intervals is f increasing?

b) On what intervals is f decreasing?

c) On what intervals is f concave up?

d) On what intervals is f concave down?

e) At what points does f have local maxima?

f) At what points does f have local minima?

g) Suppose $f(0) = 2$. What is the equation of the line tangent to f at $x = 0$?

G-3   17. The graphs of two functions are given below. One of the functions is the derivative of the other. Label the two graphs with f and f'. Explain why.

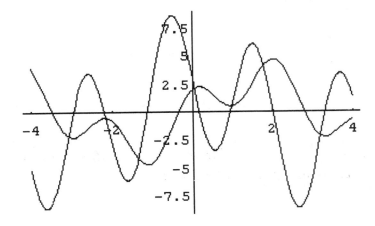

G-1   18. The following graph shows two functions. One is the derivative of the other. In other words, one is an amount function, and the other is its associated rate function. Label the two curves f and f' to show which is the amount function and which is the rate function. You should be able to give reasons why you labeled the functions as you did.

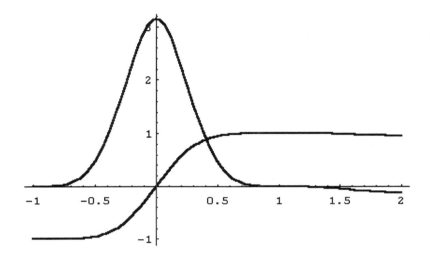

S-2   19. Suppose that f is a function whose derivative is given by:

$$f'(x) = \frac{(x-1)(x-4)^3}{1+x^4}$$

(Note: In this problem, do not try to find a formula for $f(x)$.)

a) On what interval(s) is f increasing?

b) Find all stationary points of f, and say whether each is a local maximum, a local minimum, or neither.

c) What is the slope of the tangent line to the curve $y = f(x)$ at $x = 2$?

SECTION 4: THE GEOMETRY OF HIGHER-ORDER DERIVATIVES

G-2   20. Let f be the function shown below. (The function f has a local maximum at $x = 1.14$, a point of inflection at $x = 4.67$, and a local minimum at $x = 8.19$.) Over which intervals, if any, is f' decreasing?

16

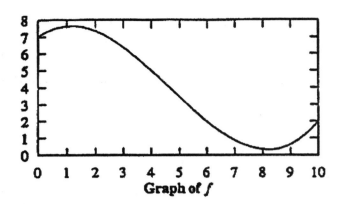

Graph of f

G-2 21. On which of the following intervals are both $\dfrac{dy}{dx} > 0$ and $\dfrac{d^2 y}{dx^2} < 0$?

    I. $a < x < b$ II. $b < x < c$ III. $c < x < d$

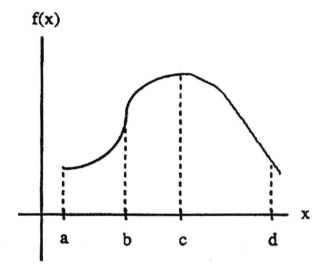

a) I only
b) II only
c) III only
d) I and II
e) II and III

G-2 22. Using technology, graph the function $f(x) = x^4 - 100x^2$. Use the graph to answer the following questions. Be careful, the graph may be deceiving. Be sure you adjust the viewing window to see all of the interesting parts of f.

    a) Estimate all zeros of f.

17

b) Estimate the positions and values of all relative extrema.

c) Estimate all points of inflection.

d) Give intervals where f is increasing and decreasing.

e) Give intervals where f is concave up and down.

G-2   23. The graph of a function g appears below. Note: g has roots at $x \approx -1.79$, $x \approx -0.13$, $x \approx 1.11$, and $x \approx 3.81$; g has stationary points at $x \approx -1.17$, $x \approx 0.52$, and $x \approx 2.90$; g has inflection points at $x \approx -0.43$ and $x \approx 1.93$.

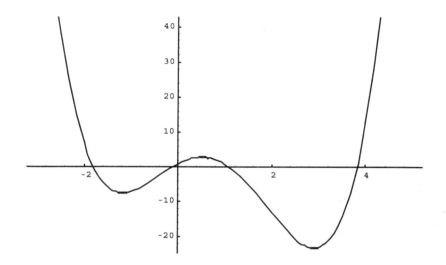

a) On what intervals is g' positive?

b) On what intervals is g' negative?

c) On what intervals is g' increasing?

d) On what intervals is g' decreasing?

e) On what intervals is g'' positive?

f) On what intervals is g'' negative?

G-2   24. Two graphs are shown below. One is a function f and the other is its second derivative f''. Tell which one is f and which one is f''.

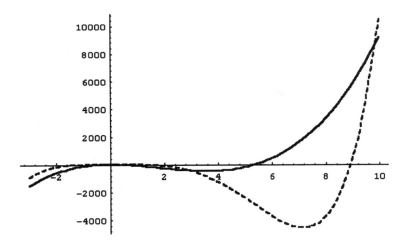

G-2   25. The graph of the second derivative of a function g is shown below. Use this graph to answer questions about g and g'. Here is the graph of g'':

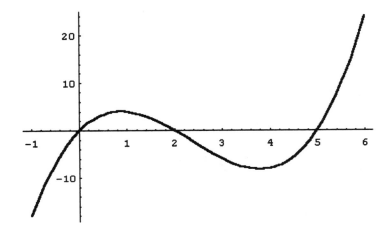

a) Where does g have points of inflection?

b) On what intervals is g concave down?

c) On what intervals is g' increasing?

G-3   26. Below is a graph showing three functions: f, f', and f''. Identify the three functions by labeling them. Explain how you arrived at your answers.

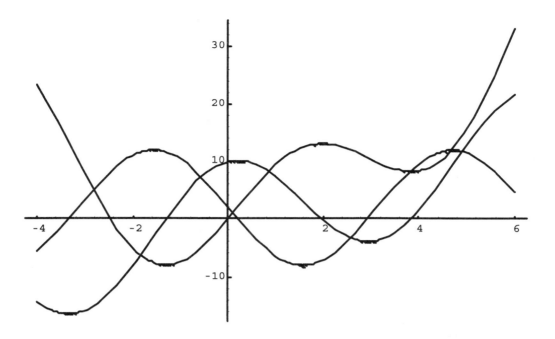

G-3   27. The graph of the derivative of a function f is given below:

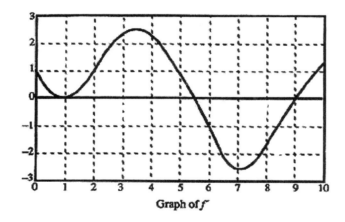

Graph of f'

a) Where does f have local maxima?

b) Where does f have local minima?

c) In what interval(s) is f concave up?

d) In what interval(s) is f concave down?

e) Suppose f passes through the point $(6, 2)$. Find the equation of the tangent line
    to the graph of f when $x = 6$.

f) Estimate the value of $f''(9)$

g) In what interval(s) is f decreasing?

h) In what interval(s) is f'' positive?

SECTION 5: AVERAGE AND INSTANTANEOUS RATES: DEFINING THE DERIVATIVE

S-2 28. Evaluate the limit by recognizing it as a derivative: $\displaystyle\lim_{h\to 0}\frac{\cos\left(\frac{\pi}{2}+h\right)-\cos\left(\frac{\pi}{2}\right)}{h}$.

S-2 29. Let k be a constant and $f(x)$ be a differentiable function. If $g(x) = f(x) + k$, then $g'(x) = f'(x)$.

a) Prove this result using the definition of the derivative.

b) Explain this result geometrically.

N-2 30. The position of a bug at time t (in minutes) along a line marked in centimeters is given by $b(t) = t^3 + 1$.

a) Find the average velocity of the bug between times $t = 2$ and $t = 4$. Be sure to include units.

b) Use technology to graph $b(t)$. Using a zoom feature, calculate the average velocity over smaller and smaller intervals. Use these calculations to find the instantaneous velocity at $t = 2$. Show the points you used and include units.

N-1 31. In the aftermath of a car accident it is concluded that one driver slowed to a halt in 6 seconds while skidding 300 feet. If the speed limit is 30 mph, can it be proven that the driver had been speeding? Explain why you are sure. (Hint: 30 mph is equal to 44 feet per second.)

N-2 32. The position $p(t)$ of an object at time t in seconds along a line marked in meters is given by $p(t) = 3t^2 - 5$.

a) Find the average velocity between $t = 2$ and $t = 4$.

b) Find the average velocity between $t = 2$ and $t = 2 + h$.

c) Find the instantaneous velocity at $t = 2$.

SECTION 6: LIMITS AND CONTINUITY

N-2   33. Evaluate the following limits. Give the unique number, ∞, $-\infty$, or D.N.E.

     a) $\lim\limits_{x \to -2^-} [x]$

     b) $\lim\limits_{x \to 3^+} [x]$

S-3   34. Evaluate the limit giving the unique number, ∞, $-\infty$, or D.N.E. as your answer.

$$\lim_{x \to 2^-} \frac{|x-2|}{x-2}$$

S-1   35. Determine if the function $h(x)$ defined below is continuous for all values in the domain. Show work to verify your conclusions.

$$h(x) = \begin{cases} x^2 + 4, & x \le 2 \\ -x^3, & 2 < x < 5 \\ -10x - 75, & x \ge 5 \end{cases}$$

G-1   36. Let f be the function whose graph is shown below. Using the graph, evaluate the following limits.

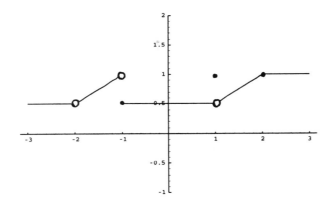

     a) $\lim\limits_{x \to 0} f(x)$

     b) $\lim\limits_{x \to 1^+} f(x)$

     c) $\lim\limits_{x \to 1^-} f(x)$

d) $\lim_{x \to 1} f(x)$

e) $\lim_{x \to -1^+} f(x)$

f) $\lim_{x \to -1^-} f(x)$

g) $\lim_{x \to -1} f(x)$

h) $\lim_{x \to -2} f(x)$

i) Where is f continuous?

S-1   37. If $\lim_{x \to 2^+} f(x) = 4$ and $\lim_{x \to 2^-} f(x) = 4$, then $\lim_{x \to 2} f(x) = $ _____ .

S-1   38. If $\lim_{x \to 2^+} g(x) = 4$ and $\lim_{x \to 2^-} g(x) = 7$, then $\lim_{x \to 2} g(x) = $ _____ .

S-2   39. a) Let $\delta = \varepsilon/3$. Show that if $|x - 2| < \delta$, then $|(3x + 1) - 7| < \varepsilon$.

b) The computation in part a) shows that $\lim_{x \to a} f(x) = L$ for some function f and numbers a and L. What is f, and what are the numbers?

$f(x) = $ _____ $a = $ _____ $L = $ _____

SECTION 7: LIMITS INVOLVING INFINITY; NEW LIMITS FROM OLD

S-2   40. Suppose that the average rate of change of a function f over the interval from $x = 3$ to $x = 3 + h$ is given by $5\,e^h - 4\cos(2h)$. What is $f'(3)$?

G-3   41. Sketch the graph of a function f on $(-\infty, -1) \cup (-1, \infty)$ such that all of the following conditions are satisfied. If it is not possible to do so, explain why.
$x = -1$ is a vertical asymptote for the graph of f.
$y = 4$ is a horizontal asymptote for the graph of f.
$f(1) = 3$, $f'(x)$ is always greater than zero and $f''(x)$ is always less than zero.

S-2   42. Evaluate the limit giving the unique number, ∞, $-\infty$, or D.N.E. as your answer.

$$\lim_{x \to \infty} \frac{5x^2 - 2 + x^4}{7x - 4x^4}$$

S-2   43. Evaluate the limit giving the unique number, ∞, - ∞, or D.N.E. as your answer.

$$\lim_{x \to 3} \frac{-5}{x - 3}$$

S-1   44. The line $y = 2$ is a horizontal asymptote of the function $f(x) = \dfrac{e^x + 2}{e^{2x} + 1}$. What limit proves this fact?

G-2   45. Sketch the graph of a function g that has all of the following properties if possible. If not possible, explain why not.

   a) g has domain $(-1, \infty)$
   b) g has range $(-\infty, 5)$
   c) $g(0) = 0$, $g(2) = 0$, $g(4) = 0$
   d) $\lim\limits_{x \to -1^+} g(x) = -\infty$
   e) $\lim\limits_{x \to 3} g(x) = -5$
   f) $\lim\limits_{x \to \infty} g(x) = 5$

N-2   46. Determine the value that best completes the following statement:

   If $\lim\limits_{x \to \infty} \dfrac{e^x}{x^p} = +\infty$, then $\lim\limits_{x \to \infty} \dfrac{x^p}{e^x} = $ _____ .

N-2   47. Let $f(x) = \dfrac{3x^2 - 11x + A}{x^2 - 9x + 20}$.

   a) For which values of A will $\lim\limits_{x \to 3} f(x)$ exist?

   b) For which values of A will $\lim\limits_{x \to 5} f(x)$ exist?

   c) For which values of A will $\lim\limits_{x \to \infty} f(x)$ exist?

CHAPTER 3: DERIVATIVES OF ELEMENTARY FUNCTIONS

SECTION 1: DERIVATIVES OF POWER FUNCTIONS AND POLYNOMIALS

Let f be the function shown below, F be an antiderivative of f, and suppose that $F(0) = 10$. (The function f has a local maximum at $x = 1.14$, a point of inflection at $x = 4.67$, and a local minimum at $x = 8.19$.)

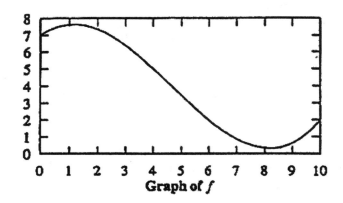

Graph of f

Questions 1 - 3 all refer to the above graph.

G-1   1. What is the slope of the F-graph at $x = 6$?

G-2   2. Is F concave down at $x = 6$?

G-2   3. Suppose $F(8) = 16$. Can $F(10) = 23$? Explain.

S-1   4. Find an antiderivative of $f(x) = (\sin x + \cos x)^2 - 2\sin x \cos x$.

S-3   5. Determine the function $f(x)$ if $f''(x) = 12x^2 - 6$ and the equation of the tangent line to the graph of f at $x = 1$ is $y = -2x + 4$.

G-3   6. Suppose f and g are functions related by the equation $g(x) = f(x - 4)$ for all x.

        a) Explain, with words and pictures, why f' and g' are then related by $g'(x) = f'(x - 4)$.

b) Use part a) to compute $g'(x)$ if $g(x) = \dfrac{1}{(x-4)^3}$.

SECTION 2: USING DERIVATIVE AND ANTIDERIVATIVE FORMULAS

S-3   7. You are driving at 30 mph (44 feet per second) when you see a child run into the road 100 feet ahead of you. You apply the brakes, giving yourself a constant deceleration of 4 feet per second. Do you stop in time? (Hint: You should find formulas expressing your velocity and your distance (from the time you applied your brakes) at any given time.)

S-2   8. A crate, open at the top, has four vertical sides, a square bottom, and a volume of 4 cubic meters. If the crate has the least possible surface area, find its dimensions.

S-3   9. A can is to be manufactured in the shape of a right circular cylinder and the volume of the can is to be 50 cubic centimeters. Find the dimensions which will minimize the total amount of material used to make the can (top, bottom and sides). Disregard any waste.

S-3   10. The metal used to make both the top and bottom of a cylindrical juice can costs 4 cents per square inch. The metal used on the sides costs 2 cents per square inch. The volume of the can is 108π cubic inches. What should the dimensions be to minimize the cost?

S-3   11. A rectangular box with square base and vertical sides is to be made to contain 640 cubic feet. The material for the base costs 35 cents per square foot, for the top costs 15 cents per square foot, and for the sides costs 20 cents per square foot. Find the dimensions of the box that will minimize cost.

S-3   12. An object is formed by adjoining a hemisphere to each end of a right circular cylinder. The total volume of the object is fixed at 12π cubic units. Find the dimensions of such an object if its surface area is to be minimized.

S-3   13. a) Find the dimensions of the rectangle with largest area that can be inscribed in an isosceles triangle with two sides of 10 inches each and one side of 6 inches. (Optional Hint: Consider similar triangles)

        b) Repeat part (a) if the triangle is equilateral with each side 10 inches.

S-3   14. A cylindrical can with closed top and bottom is to be made from two types of material: the material for the top and bottom costs $.06 per square inch and the material used for the rest of the can costs $.10 per square inch. The total cost of the can is to be \$1.44. If r is the radius and h is the height of the can, find the value for r which maximizes the volume of the can.

SECTION 3: DERIVATIVES OF EXPONENTIAL AND LOGARITHM FUNCTIONS

S-1   15. Find the derivative of $f(x) = \log_3 x$.

N-2   16. Let $g(x) = 20e^{kx}$. If $g(3) = 50$, find k.

S-2   17. Find real numbers a and b so that $y = e^x$ and $y = e^{2x}$ are both solutions of the differential equation $ay'' + by' + y = 0$.

S-1   18. Calculate the derivative of the function $f(x) = 3\ln x - \log_5 x + \ln 4$.

S-2   19. Find the antiderivative of $f(x) = \dfrac{x^2}{x^3 + 1}$.

SECTION 4: DERIVATIVES OF TRIGONOMETRIC FUNCTIONS

S-3   20. Find the derivative: $\dfrac{d}{dt}\left[\dfrac{\cos\left(2t^3\right)}{e^t + 4}\right]$.

S-2   21. Let $f(x) = e^x + 5^x + \pi^x + e^\pi - \ln x - \cos x$.

   a) Find $f'(x)$.

   b) Find $f''(x)$.

   c) Find an antiderivative $F(x)$.

S-2   22. Find an algebraic expression for $g(x)$ if $g'(x) = \cos x - 10^x + \ln 3 - \pi x^2$.

S-3   23. Calculate the derivative and antiderivative of the function
   $f(x) = e + 2e^x + 2^x + \sin x - 3\cos x$.

S-3   24. Find antiderivatives of the following functions:

   a) $f(x) = 4^x + \sin(3x) + \dfrac{1}{\sqrt[3]{x}} + \pi$

b) $f(x) = \sqrt{x} + e^{5x} + \cos(8x) + e^2 + 3x^{-2}$

SECTION 5: NEW DERIVATIVES FROM OLD: THE PRODUCT AND QUOTIENT RULES

S-2   25. Define $f(x) = \dfrac{x + \sin x}{\cos x}$, for $-\pi/2 < x < \pi/2$.

a) Find $f'(x)$

b) Find the equation of the tangent line to the graph of f at the point $x = 0$.

S-2   26. Find an algebraic expression for $g(x)$ if $g'(x) = 2x \sin x + x^2 \cos x$.

S-2   27. Let $h(x) = f(x) \cdot g(x)$ and $m(x) = f(x)/g(x)$.

x	$f(x)$	$g(x)$	$f'(x)$	$g'(x)$
-4	1	2	-2	-1
2	-3	5	4	2

Find the numerical values of the following using the entries in the table above:

a) $h'(-4)$

b) $h'(2)$

c) $m'(-4)$

d) $m'(2)$

S-2   28. Let $f(x) = e^x \sin x$. Find the exact value of x in the interval $[0, \pi]$ for which f achieves its maximum. Explain why you are guaranteed that a maximum exists.

S-2   29. Use the following graph and chart to answer this question. Note: the graph is the graph of the function $f(x)$, and the chart gives information about the function $g(x)$.

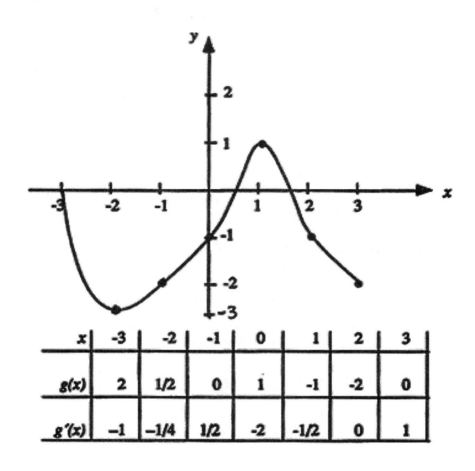

x	-3	-2	-1	0	1	2	3
g(x)	2	1/2	0	1	-1	-2	0
g'(x)	-1	-1/4	1/2	-2	-1/2	0	1

a) Let $p(x) = f(x)\,g(x)$. Find $p'(-1)$.

b) Let $q(x) = \dfrac{g(x)}{f(x)}$. Find $q'(3)$.

SECTION 6: NEW DERIVATIVES FROM OLD: THE CHAIN RULE

S-3   30. Find the derivative of $f(x) = \ln(\sin(x^2))$.

S-2   31. Find an antiderivative for $f(x) = \cos(3x) + \sin(x/3)$.

S-3   32. Some values of a function f and its derivative f' are given below:

x	−3	−2	−1	0	1	2	3	4
f(x)	4	−1	1	3	−2	2	5	7
f'(x)	0	−4	2	−1	3	−2	−3	−2

a) Let $g(x) = 3f(2x)$. Evaluate $g'(-1)$.

b) Let $h(x) = f(x^2)$. Is h increasing or decreasing at $x = 2$? Justify your conclusion.

S-2 33. Find an antiderivative of $\dfrac{1}{2x^2} + \dfrac{1}{x} + 2\sin x + e^{3x} + e^{\pi}$.

S-3 34. Complete the following table:

	$f(x)$	$f'(x)$	$f''(x)$
a)		$(x-2)^{53}$	
b)		$6(x+4)^{-\frac{5}{3}}$	
c)		$3\sin(x+2)$	
d)		e^{3x}	
e)	XXXXXXX	$\ln(2x)$	

S-3 35. Find $\dfrac{dy}{dx}$ if $y = \sin^2(e^{\ln x})$.

S-3 36. Find $f'(x)$ if $f(x) = x^3 e^{x-2} \sin(x+7)$.

S-3 37. Calculate the derivative of the given function:

a) $e^{3x}\cos(4x)$

b) $-\dfrac{7\csc(2x)}{x^{5/3} - x^4}$

c) $\sin^2(x\cos x)$

S-3 38. Find the dimensions of the rectangle of greatest perimeter that can be inscribed in a semicircle of unit radius.

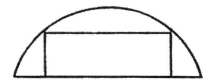

S-3   39. Let $f(x)$ be a function for which $f'(x) > 0$ for all real numbers x.

    a) Let $h(x) = f(x^2 + x)$. For what values of x is $h(x)$ increasing?

    b) Let $k(x) = f\big(f(x+3)\big)$. For what values of x is $k(x)$ increasing?

S-2   40. Use the following graph and chart to answer this question. Note: the graph is the graph of the function $f(x)$, and the chart gives information about the function $g(x)$. Let $c(x) = f(g(x))$. Find $c'(2)$.

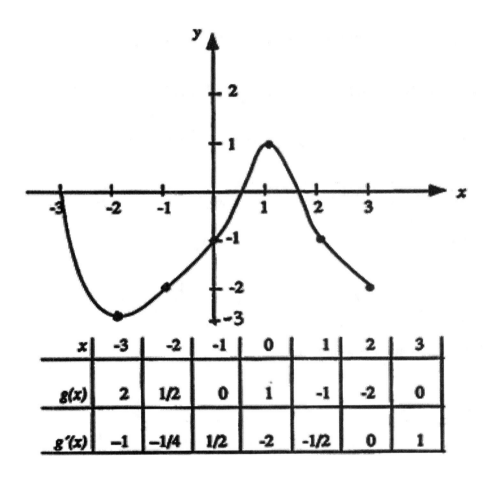

x	-3	-2	-1	0	1	2	3
$g(x)$	2	1/2	0	1	-1	-2	0
$g'(x)$	-1	-1/4	1/2	-2	-1/2	0	1

SECTION 7: IMPLICIT DIFFERENTIATION

S-2   41. Find an equation of the tangent line to the curve $3x^2 + 5y^3 = -2$ at the point $(1,-1)$.

S-3 42. Given $x^2 y = 3y + x \ln(y)$, find $\dfrac{dy}{dx}$ when $x = \sqrt{3}$.

S-3 43. Explain why (3, 1) is on the graph of the equation $x^2 y + \cos(y-1) = 10$ and then find the slope of this graph at the point (3, 1).

S-2 44. Assume $2x^3 + 3xy + y^3 = 2$ defines $y = f(x)$ implicitly. Find $\dfrac{dy}{dx}$.

S-2 45. Find the equation of the tangent line to the curve $x^2 + xy - y^3 = xy^2$ at the point (1,1).

S-3 46. Find the slope of the tangent line to the curve $(1+xy)^3 = 3y + 5$ at the point (1, 1).

SECTION 8: INVERSE TRIGONOMETRIC FUNCTIONS AND THEIR DERIVATIVES

S-3 47. Calculate the derivative of the function $f(x) = \arctan \sqrt{x-1}$.

S-2 48. Find the antiderivative of $f(x) = \dfrac{(\arctan x)^3}{1+x^2}$.

S-3 49. Let $f(x) = (\arctan(3x))^2$. Find $f'(\pi/12)$.

S-3 50. Find the antiderivative of the function $w = \dfrac{1+\tan^2 v}{\sqrt{1-\tan^2 v}}$.

S-2 51. Differentiate $y = e^x \arcsin x$.

CHAPTER 4: APPLICATIONS OF THE DERIVATIVE

SECTION 1: DIFFERENTIAL EQUATIONS AND THEIR SOLUTIONS

S-2   1.   Is the function $y(x) = \sqrt{4x+3}$ a solution of the differential equation $y \cdot y' = 2$ over the interval $(-1,1)$? Explain.

S-3   2.   A new coffee cup has constant of proportionality $k = -.02\,\text{min}^{-1}$. If the outside temperature is $10°$ C and the coffee starts at $90°$ C:

      a)   How long does the coffee remain above $75°$ C in the cup?

      b)   How hot is the coffee after 20 minutes?

      c)   At what rate is the temperature changing 4 minutes after the coffee was purchased?

S-3   3.   A detective discovers a dead body outside at 7 am and, upon measuring its temperature, finds it to be $28°$ C. An hour and a half later, the temperature is measured again and found to be $24°$ C. The meteorological office confirms that the temperature has been $15°$ C throughout the night and early morning. Estimate when the death occurred. (Hint: A healthy person's normal body temperature is $37°$ C.) Assume that the rate of change of temperature with respect to time is proportional to the difference between the body's temperature and the external temperature.

S-2   4.   Consider a cup of hot coffee cooling in a room. This situation can be modeled by Newton's Law of Cooling, $y' = k(y - T)$, where y is the temperature of the coffee, T is the temperature of the room and k is a constant. Is the constant k positive or negative? Explain your answer.

S-3   5.   Consider a ball falling from some height. It can be shown that under certain assumptions, this situation can be modeled by the initial value problem $mv' = -mg - kv$, with $v(0) = v_0$, where m is the mass of the ball, v is the velocity of the ball, g is the force of gravity, and k is a constant. Verify that
$$v(t) = -\frac{mg}{k} + \left(v_0 + \frac{mg}{k}\right)e^{-kt/m} \text{ is a solution of the initial value problem.}$$

SECTION 2: MORE DIFFERENTIAL EQUATIONS: MODELING GROWTH

N-1   6. If you, at age 18, want to put away some money so you can retire at the age of 70 with $2 million, how much money do you need to deposit, assuming a nominal interest rate of 2%? Assume that the rate at which the amount grows is proportional to the amount present.

S-2   7. a) Suppose that the growth of a certain biological population can be modeled by the exponential growth equation $P'(t) = .05\ P(t)$, with $P(0) = P_0$. Note that this model does **not** include any consideration of mortality. The growth is unrestricted. Suppose that the death rate is also proportional to the current population. Using a constant of proportionality d (for "death"), modify the differential equation so that it contains separate terms for births and deaths.

   b) Immigration laws permit 1,000,000 people per year to enter the country (independent of the current population). Modify your answer to (a) to include this additional information.

S-3   8. A natural gas leak has filled a 100,000 cubic meter building with a 1 percent mixture of natural gas and air. The gas line is shut off and an emergency ventilation system pumps in fresh air at the rate of 2,000 cubic meters per minute. Assume that the gas and air mix instantaneously and uniformly, and that the mixture vents at the same rate that fresh air is introduced.

   a) Explain why $v(t)$, the volume of gas in the building at time t, is the solution of the IVP

   $$v' = -.02\,v \ \text{ with } v(0) = 1,000$$

   b) Using Theorem 1, page 290, find a solution to this IVP.

   c) How much gas is left in the building after 1 hour? (Optional Hint: t is in minutes.)

S-2   9. Doubling time is defined to be the length of time necessary for a given quantity to double in size. Recall that money invested at a certain interest rate r which is compounded continuously is modeled by the differential equation $y' = ry$, with $y(0) = y_0$ as the initial amount invested. Consider a credit card with a balance of $5,000 and an interest rate of 17.9% compounded continuously. The balance satisfies the differential equation given above.

   a) How long will it take until the balance doubles if no payments are made? (In other words, find the length of time until the balance is $10,000.)

   b) How long until the balance grows from $10,000 to $20,000?

    c) From $20,000 to $40,000?

    d) What do you notice about the answers to parts (a), (b), and (c) ?

    e) What does this tell you about doubling time?

S-1   10. Recall that the formula for interest compounded n times per year is given by

$$A(t) = P_0(1 + \tfrac{i}{n})^{tn}$$

where $A(t)$ is the amount of money present at time t, P_0 is the initial investment, t is the time in years and i is the interest rate. Consider an initial investment of $50,000 at an interest rate of 8%. How much money will there be after 10 years if interest is compounded:

    a) once a year

    b) once a month

    c) once a day

    d) once an hour

    e) once a minute

    f) continuously

    g) How are the answers in (a) - (e) related to the answer in (f)?

SECTION 3: LINEAR AND QUADRATIC APPROXIMATION; TAYLOR POLYNOMIALS

S-2   11. Suppose that f is a function such that $f'(x) = e^{\sin x}$ and $f(0) = 1$. (Note: no explicit formula for f is given.)

    a) Estimate the value of $f(.25)$ using a linear approximation.

    b) Estimate the value of $f(.25)$ using a quadratic approximation.

S-3   12. Find the third degree Taylor Polynomial for $f(x) = \sin(2x)$ centered at $x = 0$. For what values of x does this polynomial appear to be a good approximation for $\sin(2x)$ (accurate to within $\pm.01$)? Note: Answers may vary.

S-2   13. Find the Taylor polynomial of degree 2 for $f(x) = x^3 - 3x^2 + 3$ centered at $a = 1$.

S-2   14. Let $y = \sqrt{x}$. Find the fifth degree Taylor polynomial approximation for this function at the value $x_0 = 1$.

S-2   15. Consider the Taylor polynomial of order 4, based at x_0 given below. Find the function f which it approximates and find x_0.

$$P_4(x) = e^2 + 2e^2(x-1) + 2e^2(x-1)^2 + \tfrac{4}{3}e^2(x-1)^3 + \tfrac{2}{3}e^2(x-1)^4$$

S-2   16. a) Calculate the third degree Taylor polynomial $p(x)$ which has the same value and the same first three derivatives as $f(x) = x \sin x$ at $x = \pi/2$.

          b) Compare the behavior of $p(x)$ and $f(x)$ as x approaches ∞.

S-2   17. a) Find the Taylor polynomial $P_1(x)$ of degree 1 for the function $f(x) = x + e^x$ based at 0.

          b) Using technology, graph f and P_1 for $0 < x < 1$. Is the graph of f above or below the graph of P_1? Justify your answer by finding the sign of $f(x) - P_1(x)$; the graph is not sufficient.

          c) Find the Taylor polynomial $P_2(x)$ of degree 2 for the function $f(x) = x + e^x$ based at 0.

          d) Find the Taylor polynomial of degree 2 for the function $f(x) = x + e^x$ based at 1.

S-2   18. Find the quadratic approximation to $f(x) = xe^x$ based at 0.

SECTION 4: NEWTON'S METHOD: FINDING ROOTS

G-2   19. To find the value of the root r, Newton's method is used, with a starting value $x = c_0$.

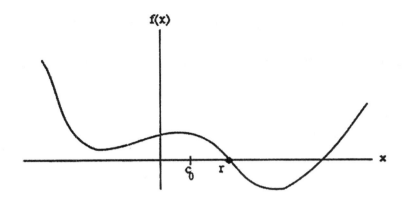

a) Show the next two approximations, c_1 and c_2 on the graph.

b) Explain what will happen as we continue the process.

G,N-1 20. a) Apply Newton's method to $f(x) = \cos x$ starting with $x_0 = \pi/4$ to approximate a root of the function. Calculate x_1, x_2, and x_3. What is the general pattern of the estimates?

b) Sketch a graph of the function f and the associated tangent lines to the graph of f at $x = x_0$, x_1, and x_2 to demonstrate what is happening geometrically in this root approximation process.

N-2   21. Write down the formula for x_{n+1} in Newton's Method. Use technology (and Newton's method) to find the largest root of $f(x) = x^3 - 3x + 1$ accurate to 6 decimal places. (Optional Hint: The largest root is in $[1, 2]$.)

G-2   22. Consider the graph of a function $f(x)$ given below.

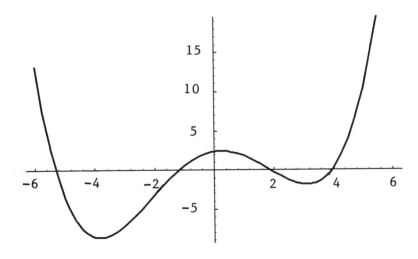

a) On the graph carefully draw the line that is tangent to the graph of f at $x = 3$.

b) Circle the root to which Newton's Method will converge if you take $x = 3$ as the starting value.

c) What geometric property of the tangent line drawn in part (a) causes Newton's Method to converge to the "wrong" root?

S-3 23. Explain how Newton's method can be used to investigate solutions to the equation $e^{-x} = 10\ln x$ by defining an appropriate function $f(x)$ and carrying out one iteration starting with $x_0 = 1$.

S-3 24. Use Newton's method to find the solution of the equation $2x = \tan x$ in the interval $0 < x < \pi/2$ correct to 8 decimal places.

SECTION 5: SPLINES: CONNECTING THE DOTS

S-3 25. Find the quadratic spline S, through the three knots $(0, 2)$, $(1, -3)$, and $(2, 0)$. Assume that $S'(0) = 2$.

S-3 26. Construct a cubic spline S through the knots:

$(1, 3)$, $(2, 2)$, $(3, 1)$ with $S'(1) = 1$, $S''(1) = 0$

S-3 27. Find constants a, b, and c so that the function given below is a quadratic spline through the knots $(1, 1)$, $(2, 5)$ and $(3, -1)$.

$$S(x) = \begin{cases} S_1(x) = a + b(x-1) + c(x-1)^2, & \text{if } 1 \le x \le 2 \\ S_2(x) = 5 + 9(x-2) - 15(x-2)^2, & \text{if } 2 \le x \le 3 \end{cases}$$

SECTION 6: OPTIMIZATION

S-2 28. A crate open at the top has vertical sides, a square bottom, and a volume of 4 cubic meters. If the crate has the least possible surface area, find its dimensions.

S-2 29. A right circular cone with height h and base radius r has volume $V = \frac{1}{3}\pi r^2 h$. Find the maximum volume of the right circular cone given that it s dimensions satisfy $6r + h = 18$. Also find the height and radius of the cone with maximum volume. Use the second derivative test to justify your conclusion.

S-2   30. Find the largest possible area for a rectangle with base on the x - axis and upper vertices on the graph of $y = 5 - x^2$.

S-3   31. An island is at point A, 4 miles offshore from the nearest point B on a straight beach. A store is at point C, 7 miles down the beach from point B. If a person can row at the rate of 2 mph and walk at the rate of 5 mph, where should a person land in order to go from the island to the store in the least possible time?

S-2   32. A printing company has twelve presses, each of which can print 1000 copies per hour. It costs $5 to set up each press for a run and $100 + 5n$ dollars to run n presses for one hour. Determine how many presses should be used in order to print 5000 copies of a poster at the lowest possible cost. Be sure to verify your answer to be sure that you have in fact obtained a minimum.

SECTION 7: CALCULUS FOR MONEY: DERIVATIVES IN ECONOMICS

S-2   33. The market research department of a company recommends that the company manufacture and market a new transistor radio. After the test marketing, the research department presents the following cost and revenue equations:

$$C(x) = 7,000 + 2x$$
$$R(x) = 10x - \frac{x^2}{1,000}$$

where x is the number of radios produced.

a) Find the marginal cost.

b) Find the marginal revenue.

c) Find an equation for profit.

d) Find the marginal profit.

e) Find the break-even points (the points where revenue equals cost).

f) What level of production will maximize profit?

g) What is the maximum profit?

h) How much should the selling price be to maximize profit?

S-2   34. The Sitting Pretty Chair Company can manufacture a certain style chair for $50 each. If they sell them for $80 each, they can sell 500 chairs a month. For every $2 they reduce the price, the number of sales increases by 100.

   a) What price maximizes profit?

   b) How much profit is made at this price?

S-2   35. The cost of printing q copies of a calculus text is $1000 + 3q$ dollars. If the selling price is p, the number of copies sold will be $1000 - 8p$.

   a) What price maximizes profit?

   b) How much profit is made at this price?

   c) How many books will be sold?

S-2   36. The cost of manufacturing q double handled coffee cups (for ambidextrous people) per week is

$$C(q) = 100 + 20q + q^3$$

What is the smallest selling price for which it would make sense to manufacture any cups? (Optional Hint: If p is the price, set revenue $R(q) = pq$ equal to cost $C(q)$, solve for p and minimize.)

S-2   37. Pizza World can sell 90 pizzas a day if it charges $5 for a pizza. For every $1 increase in price, Pizza World sells 20 fewer pizzas a day. It costs the restaurant $3 to make each pizza. Find the price Pizza World should charge for its pizzas to maximize its profit.

SECTION 8: RELATED RATES

S-2   38. The sides of the rectangle below increase in such a way that $\dfrac{dz}{dt} = 1$ and $\dfrac{dx}{dt} = 3\dfrac{dy}{dt}$.

At the instant when $x = 4$ and $y = 3$, what is the value of $\dfrac{dx}{dt}$?

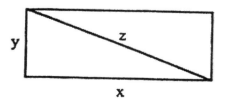

40

S-3 39. You are riding a Ferris Wheel 120 ft. in diameter at the Indiana State Fair. It makes one complete revolution per minute. How fast are you falling when you're halfway to the bottom? (Hint: Express your height in terms of the angle your spoke makes with the horizontal; make the center of the Ferris wheel the origin.)

S-2 40. A boat is being pulled into a dock through a pulley on the dock that is 20 ft. above the deck of the boat. If the rope is drawn in at a rate of 6 ft./min., how fast is the boat approaching the dock when it is 15 feet away from the dock?

S-3 41. A baseball diamond is a square with sides 90 feet long. Suppose a baseball player is advancing from second to third base at the rate of 24 feet per second, and an umpire is standing on home plate. Let θ be the angle between the third baseline and the line of sight from the umpire to the runner (see the figure below). How fast is θ changing when the runner is 30 feet from third base?

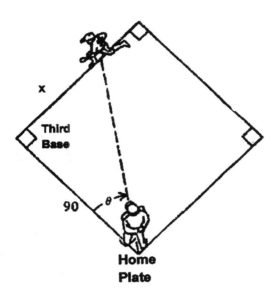

S-3 42. A swimming pool is 20 feet wide, 40 feet long, and 10 feet deep at one end. The depth gradually decreases to 0 at the other end. Water is being added to the pool at the rate of 16 cubic feet per minute. Find the rate at which the water level is rising when there is 400 cubic feet of water in the pool.

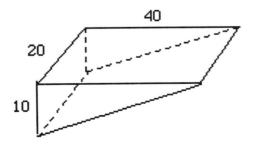

SECTION 9: PARAMETRIC EQUATIONS, PARAMETRIC CURVES

S-3   43. If $x(t) = t^2$ and $y(t) = t^4$,

        a) Write the curve in a form involving only x and y.

        b) Plot the parametric curve on $[-1,1]$, labeling $t = -1, 0, 1$.

        c) Is this a smooth curve for all t in $[-1,1]$? Why or why not?

G-2   44. a) Sketch the parametric curve:

$$x = 4(1 - \sin t)$$
$$y = 4(1 - \cos t)$$

        for $0 \le t \le 2\pi$.

        b) Mark the direction of travel and label the points corresponding to $t = 0, t = \pi,$ and $t = 2\pi$.

        c) What is the slope of the curve at $t = \pi/4$?

S-2   45. Consider the parametric equations

$$x = \sin(\pi t)$$
$$y = \cos(2\pi t)$$

        for $0 \le t \le 2$.

        a) Is it possible to eliminate the parameter t in the two equations above and find a single equation in x and y for the curve? (Using technology to plot the parametric equations may help you decide.)

        b) If you answered "yes" to part (a), eliminate t and find the single equation.

S-3   46. Consider the parametric equations

$$x = \sin t + 2$$
$$y = \cos t - 2$$

        for $0 \le t \le 2$.

    a) Using technology sketch the graph of the equations and indicate the direction of travel as t increases.

    b) What happens to the graph if the bounds on t change to $-2 \leq t \leq 4$? Sketch this graph.

    c) Does the direction of travel change if the bounds on t change? Explain.

    d) Can the parameter be eliminated? If so, find a single equation in x and y for the curve.

S-3   47. Consider the parametric equations given by

$$x = 3\cos\tfrac{t}{3} - \cos t$$
$$y = 3\sin\tfrac{t}{3} - \sin t \,.$$

    a) Using technology sketch the curve represented by these equations.

    b) Find an equation of the tangent line to the curve at the point where $t = \pi/2$.

    c) Find all points of horizontal and vertical tangency.

SECTION 10: WHY CONTINUITY MATTERS

G-3   48. Determine whether the following statement **must** be true, **might** be true, or **cannot** be true. Quote the appropriate theorem and justify your conclusion. You may also use a diagram.

If f is continuous and has no roots in $[2, 7]$, then $f(2) \cdot f(7) < 0$.

N-1   49. A function f that is continuous for all real numbers x has $f(3) = -1$ and $f(7) = 1$. If $f(x) = 0$ for exactly one value of x, then which of the following could be x? Justify your answer.

    a) -1 b) 0   c) 1   d) 4   e) 9

N-1   50. State the Intermediate Value Theorem and use it to show the function $g(x) = x^3 + \cos(47x)$ must take on the value -17.1 somewhere on the interval $[-3, 0]$. Explain yourself clearly.

G-1   51. The sketch below shows part of the graph of a continuous function $f(x)$.

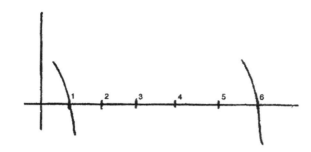

Explain why $f(x)$ must have a root in the interval (1, 6).

S-1   52. a) Use the Intermediate Value Theorem to show that $f(x) = x^5 + x^3 - 6x - 1$ has a root in [0, 2]. Explain how you are applying the theorem.

        b) Determine which half of the interval [0, 2] definitely has a root. Explain how you decided.

SECTION 11: WHY DIFFERENTIABILITY MATTERS; THE MEAN VALUE THEOREM

G-2   53. Determine whether the following statement **must** be true, **might** be true, or **cannot** be true. Quote the appropriate theorem and justify your conclusion.

        If f is continuous on [3, 7], and differentiable on (3, 7), and $f(3) = 5$ and $f(7) = 2$, there must be a number c between 3 and 7 so that $f'(c) = -3/4$. Be sure to include a diagram to illustrate your explanation.

G-2   54. Draw a sketch to illustrate what the Mean Value Theorem says for the specific function x^3 on the interval $[-1, 3]$. Also, find a specific value for the special point c guaranteed by the MVT for the given function on the given interval.

S-3   55. Carefully state the Mean Value Theorem and then use it to prove that any function f that has $f'(x) < 0$ for every x in [a, b] must have $f(b) < f(a)$.

S-1   56. Find a number c in the interval (0, 5) that satisfies the Mean Value Theorem for $f(x) = x^2 + 3x - 1$.

S-2   57. Let $g(x) = x^{\frac{2}{3}} - 1$. Notice that although $g(-1) = g(1) = 0$, there is no point c in the interval $(-1, 1)$ where $g'(c) = 0$. Explain why this doesn't contradict Rolle's Theorem.

CHAPTER 5: THE INTEGRAL

SECTION 1: AREAS AND INTEGRALS

The graph of a function $h(x)$, consisting of semicircles and straight lines, is given below. Use the graph to answer questions 1 and 2. Be sure to show all work.

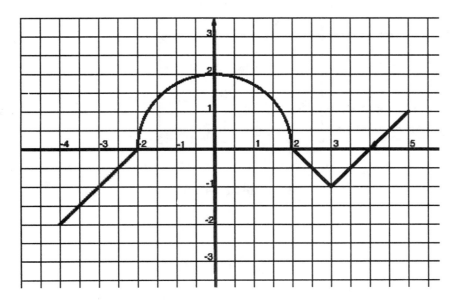

G-2   1. Calculate $\displaystyle\int_{-4}^{5} h$.

N-1   2. Calculate the average value of $h(x)$ over the interval $[-4, 5]$.

G-1   3. Show that $\displaystyle\int_{-1}^{2} g(x)dx$ is between 2 and 5. The graph of g is given below.

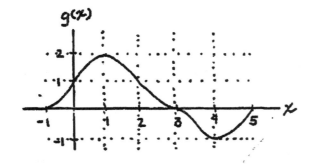

45

The graph of a function g is shown below. The graph consists of straight lines and semicircles:

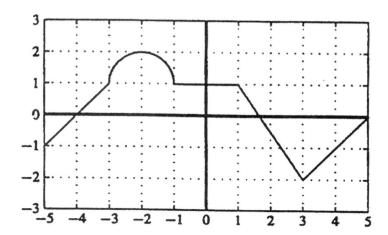

Use the graph to evaluate the integrals in problems 4 - 6.

G-2   4. $\displaystyle\int_{-5}^{0} g(x)\,dx$.

G-2   5. $\displaystyle\int_{0}^{5} 3\,g(x)\,dx$.

G-2   6. $\displaystyle\int_{-5}^{5} |g(x)|\,dx$.

S-1   7. Suppose that $\displaystyle\int_{-3}^{5} f(x)\,dx = 4,\ \int_{-3}^{5} g(x)\,dx = 5$ and $\displaystyle\int_{-3}^{5} h(x)\,dx = -3$. Using properties of the integral, find the following:

a) $\displaystyle\int_{5}^{-3} (h(x) - f(x))\,dx$

b) $\displaystyle\int_{-3}^{5} (-3g(x) + 4)\,dx$

S-2   8. Suppose that $\int_0^2 f(x)\,dx = 7,$ $\int_1^2 f(x)\,dx = -4$, and $\int_2^4 f(x)\,dx = 8$.

   a) Draw the graph of a function with all of these properties.

   b) Find $\int_1^4 f(x)\,dx$.

   c) Find $\int_0^1 f(x)\,dx$.

   d) Find $\int_0^1 f(x+1)\,dx$

   e) Explain why f must be negative somewhere in the interval $[1, 2]$.

   f) Explain why $f(x) \geq 4$ for some value of x in the interval $[2, 4]$.

SECTION 2: THE AREA FUNCTION

G-2   9. For which value of c in $[-1, 5]$ will $\int_{-1}^c g(x)\,dx$ be the largest? Justify your answer. The graph of g is given below.

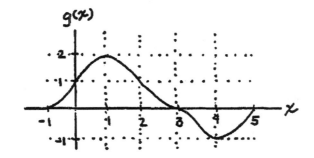

S-1   10. Consider the following functions:

$g(x) = c,$ $h(x) = x,$ $i(x) = 2x,$ $j(x) = 3x$ where c is a constant.

a) Recall that the area function for a function f is given by $A_f(x) = \int_0^x f(t)\,dt$.

Calculate the area function for each of the functions listed above.

b) Using your answers to part a), calculate $\dfrac{d}{dx}A_f(x)$ for each of the functions listed above.

c) What do you notice about the answers to part (b)?

d) Make a conjecture about the formula for calculating $\dfrac{d}{dx}A_f(x)$ for any function f.

G-1   11. The graph of an area function $A_f(x)$ is given below. Use the graph to answer questions about the function f. (Note: $A_f(x)$ has stationary points at $x \approx -0.312$ and $x \approx 1.423$ and has a point of inflection at $x \approx 0.556$)

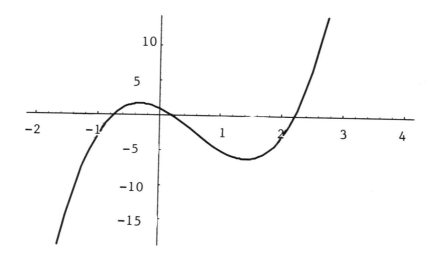

a) On which intervals is f positive?

b) On which intervals is f negative?

c) On which intervals is f increasing?

d) On which intervals is f decreasing?

e) Where is f zero?

G-1   12. The graph of an area function $A_f(x)$ is given below. Use the graph to answer questions about the function f. (Note: $A_f(x)$ has stationary points at $x \approx -0.843$, $x \approx 0.593$ and $x = 4$ and has points of inflection at $x \approx -0.186$ and $x \approx 2.686$)

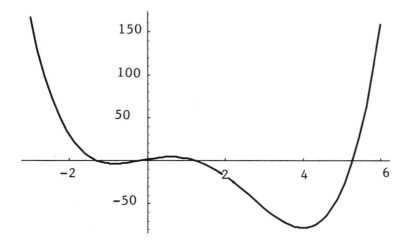

a) On which intervals is f positive?

b) On which intervals is f negative?

c) On which intervals is f increasing?

d) On which intervals is f decreasing?

e) Where is f zero?

G-2   13. The graphs of two functions are shown below. One is the graph of a function f and the other is the graph of its area function $A_f(x) = \int_0^x f(t)\,dt$. Determine which is f and which is $A_f(x)$. Explain your choices.

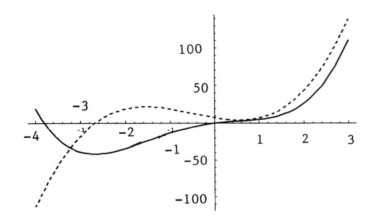

G-2   14. The graphs of two functions are shown below. One is the graph of a function f and the other is the graph of its area function $A_f(x)$. Determine which is f and which is $A_f(x)$. Explain your choices.

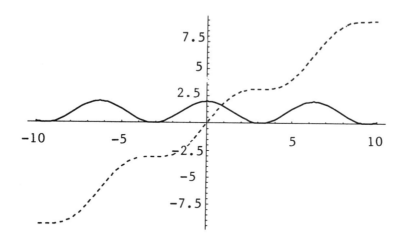

SECTION 3: THE FUNDAMENTAL THEOREM OF CALCULUS

S-2   15. If $\int\limits_{0}^{k}\left(2kx - x^2\right)dx = 18$, find k.

S-2   16. Find the derivative: $\dfrac{d}{dx}\left[\int\limits_{0}^{x}\sin\left(t^6\right)dt\right]$.

S-3   17. Let $f(x) = \begin{cases} -4, & x < 0 \\ 2, & x \geq 0 \end{cases}$, and let $F(x) = \int\limits_{-2}^{x} f(t)\,dt$.

   a) Evaluate $\int\limits_{-2}^{6} f(x)\,dx$.

   b) Evaluate $\int\limits_{7}^{4} f(x)\,dx$.

   c) Sketch the graph of $F(x)$.

   d) Find an algebraic formula for $F(x)$.

e) Find $F'(-3)$.

f) Find $F'(4)$.

S-2   18. Let f be a function with the following properties:

a) $f'(x) = ax^2 + bx$

b) $f'(1) = 12$

c) $f''(-1) = 21$

d) $\displaystyle\int_1^2 f(x)\,dx = 20$

Find an algebraic expression for $f(x)$.

S-3   19. Calculate $f'(2)$ if $f(x) = x \cdot \displaystyle\int_2^x (t^2 + 1)\sin(\pi t)\,dt$.

S-2   20. a) Calculate $f'(x)$ if $f(x) = x \cdot \sin x$.

b) Use your answer to part (a) to evaluate $\displaystyle\int_0^{2\pi} (3x\cos x + x + 3\sin x)\,dx$.

SECTION 4: APPROXIMATING SUMS: THE INTEGRAL AS A LIMIT

N-2   21. a) Find a linear approximation for $\sin(x^2 - x)$ around $x = 0$.

b) Use it to approximate $\displaystyle\int_0^1 \sin(x^2 - x)\,dx$.

c) Approximate this integral by calculating M_{100} (a midpoint approximating sum using 100 subintervals) using technology, and compare the value with your answer from part b). How far apart are the answers?

N-2   22. Let $f(x) = x + \sin x$. Estimate the value of $\int_0^3 f(x)\,dx$ by evaluating the left sum with three equal intervals. Draw a sketch that illustrates this sum geometrically, then evaluate the integral by using the Fundamental Theorem of Calculus.

S-2   23. Find a definite integral that is approximated by the left sum: $\dfrac{2}{5} \displaystyle\sum_{j=0}^{19} e^{(\frac{2}{5})j}$.

N-2   24. Consider the function $f(x) = 3x^2$ on the interval $[-2, 2]$. The graph is given below.

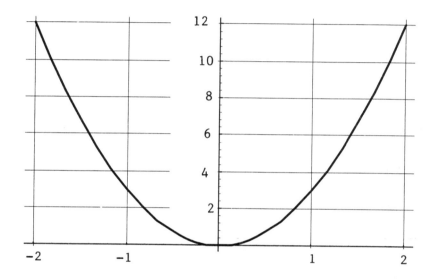

a)   Calculate $I = \int_{-2}^{2} f(x)\,dx$ exactly using the Fundamental Theorem of Calculus.

b)   Fill in the chart below. Calculate the right sums needed by using technology. You will also need to calculate absolute and relative errors. Use the formulas given here:

Absolute error = the absolute value of the difference between the estimate and the actual answer

$= |I - R_n|$, where R_n is a right approximating sum using n subintervals

Relative error = the absolute value of the ratio of absolute error to the actual answer

$= \left| \dfrac{I - R_n}{I} \right|$

n, number of subintervals	R_n	absolute error	relative error
1			
2			
4			

N-1 25. Using technology, approximate the value of $\int_0^{2\pi} \sin(t^2)\,dt$ accurate to within .01 using:

    a) left approximating sums

    b) midpoint approximating sums

SECTION 5: APPROXIMATING SUMS: INTERPRETATIONS AND APPLICATIONS

N-2 26. A device attached to a car measures the speed of the car once a minute. The measured speeds are given in the table below. Using trapezoids, approximate the total distance traveled by the car during the ten minute interval given.

time (min)	0	1	2	3	4	5	6	7	8	9	10
speed (mph)	0	70	61	93	124	117	89	56	49	37	0

N-1 27. a) Sketch the region bounded by the curves $y = x^3$, $x = 0$, and $y = 3x + 2$.

    b) Approximate the area of the region using two left rectangles.

    c) Approximate the area of the region using four left rectangles.

    d) Calculate the exact area of the region by integration.

In problems 28 - 30, sketch the region bounded by the given curves and find the area of the region exactly:

S-2 28. $y = \sin x$, $y = e^x$, $x = 0$, $x = \pi$

S-3 29. $y = \sqrt{x}$, $y = \cos x$, $x = 0$ Note: Use technology to approximate the point of intersection.

S-2 30. $y = \sin x$, $y = -1$, $x = 0$, $x = 2\pi$

CHAPTER 6: FINDING ANTIDERIVATIVES

SECTION 1: ANTIDERIVATIVES: THE IDEA

In problems 1 - 3, evaluate the antiderivative. Check your answer by differentiation.

S-1   1. $\int \left(3x + e^{3x}\right) dx$

S-2   2. $\int \left(\dfrac{1}{5x} - \dfrac{1}{5x^2} + \dfrac{1}{5}\right) dx$

S-2   3. $\int \dfrac{3\pi}{\sqrt{\pi^2 - x^2}}\, dx$

In problems 4 - 6, check the antiderivative formulas by differentiation.

S-2   4. $\int \dfrac{dx}{x(2 + 3x)} = -\dfrac{1}{2}\ln\left|\dfrac{2 + 3x}{x}\right| + C$

S-2   5. $\int \dfrac{dx}{x\sqrt{2x - 1}} = 2\tan^{-1}\sqrt{2x - 1} + C$

S-2   6.   Be sure to verify both forms of the formula. Note: You may need to use trigonometric identities to verify.

$$\int \sin^2 3x\, dx = -\tfrac{1}{6}\cos 3x \sin 3x + \tfrac{1}{2}x + C = \tfrac{1}{2}x - \tfrac{1}{12}\sin 6x + C$$

SECTION 2: ANTIDIFFERENTIATION BY SUBSTITUTION

S-2   7.   Find real numbers a and b so that the equality holds, then evaluate the definite integral.

$$\int_0^2 e^{5x}\cos\left(e^{5x}\right) dx = \tfrac{1}{5}\int_a^b \cos u\, du$$

S-3   8.   Suppose $\int_c^{2c} e^{f(x)} dx = 7$ where c is some constant. Evaluate: $\int_1^2 e^{f(cx)} dx$

S-2   9. Evaluate. Check your answer by differentiation. $\int e^{2x} \sec^2(e^{2x})\,dx$

S-2   10. Evaluate. Check your answer by differentiation. $\int \dfrac{\sin x}{\sqrt{1-\cos^2 x}}\,dx$

S-3   11. Evaluate: $\displaystyle\int_0^{\sqrt{2}/8} \dfrac{3}{\sqrt{1-16x^2}}\,dx$

S-3   12. Evaluate: $\displaystyle\int_0^{\pi/4} 2^{\tan x} \sec^2 x\,dx$

SECTION 3: INTEGRAL AIDS: TABLES AND COMPUTERS

In problems 13 - 17, evaluate the integrals using a table or technology. Making simple *u*-substitutions or completing the square first may help.

S-2   13. $\int e^{2x}\left(5e^{2x}+14\right)^2 dx$

S-2   14. $\int \dfrac{e^{2x}}{\sqrt{e^{2x}-4}}\,dx$

S-2   15. $\int \sin 2x\sqrt{3\cos 2x+1}\;dx$

S-3   16. $\int (2x-3)\sin\left(x^2-3x\right)\cos\left(2x^2-6x\right)dx$

S-3   17. $\int \dfrac{\sec^2 x}{1-\pi\,e^{2\tan x}}\,dx$

CHAPTER 7: NUMERICAL INTEGRATION

SECTION 1: THE IDEA OF APPROXIMATION

N,S-1 1. Consider $\int_{1}^{2} x^{-1}dx$. The graph of $f(x) = x^{-1} = 1/x$ is given below.

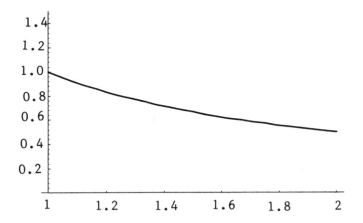

    a) Calculate the integral exactly.

    b) Using technology, find T_{10}.

    c) Using your answers from (a) and (b), calculate the actual error from using T_{10}.

    d) Estimate the error for T_{10} using the estimate for monotone functions.

G-2   2. Illustrate on the graph below that $\left| R_3 - L_3 \right| = \left| f(b) - f(a) \right| \cdot \dfrac{(b-a)}{3}$ and explain what this has to do with the error bounds formula for left and right sums.

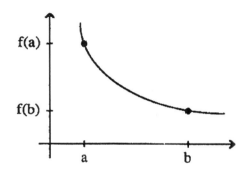

N-1   3. Calculate the right approximating sum for $f(x) = 1/x$ from $x = 1$ to $x = 4$, using 3 subintervals by hand. The graph of $f(x)$ is given below.

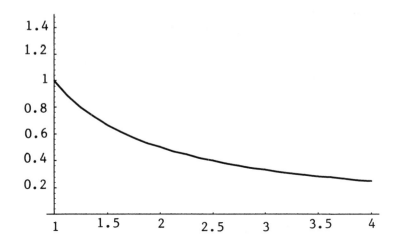

N-3   4. Consider the integral $I = \int_{0}^{5} \cos\left(\frac{\pi}{2} x\right) dx$. The graph of $f(x) = \cos\left(\frac{\pi}{2} x\right)$ is given below.

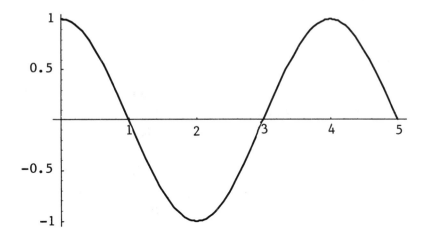

a) Compute L_5 by hand.

b) Draw the rectangles on the graph of f.

c) Estimate the error from using L_5 to approximate I using the error estimate for monotone functions. Note: Find a way to use this estimate even though f is not monotone on [0, 5].

d) Compute I and calculate the actual error.

N,S-2 5. Consider the integral $I = \int_{0}^{5} (3x)^{1/2} dx$.

    a) Use technology to estimate the value of I by using a right sum with 10 subintervals.

    b) Use summation notation to write R_{10} from (a).

N-1 6. Consider the integral $I = \int_{-1}^{2} e^{-x} dx$. The graph of $f(x) = e^{-x}$ is given below.

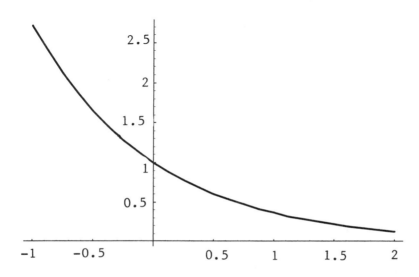

    a) Compute L_3 and R_3 by hand. You may use a scientific calculator.

    b) Use technology to calculate T_3 .

    c) Use the Fundamental Theorem of Calculus to find I exactly.

    d) Find the exact errors in using L_3 , R_3 , and T_3 .

    e) What is the relationship between the errors?

SECTION 2: MORE ON ERROR: LEFT AND RIGHT SUMS AND THE FIRST DERIVATIVE

G,N-2 7. Consider the integral $I = \int_{0}^{2} \sin(x^2) dx$. The graph of $f(x) = \sin(x^2)$ is given below.

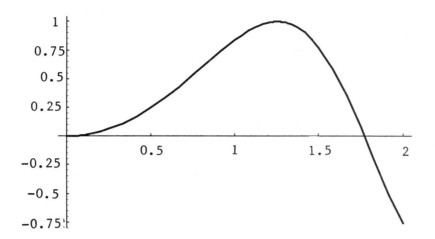

a) Compute R_4 by hand (you may use a scientific calculator).

b) Draw the four right sum rectangles on the graph of $f(x) = \sin(x^2)$ and shade them in.

c) Calculate an error bound for using R_4 to estimate the value of I.

N,S-2   8.   Let $f(x) = \sqrt{x} = x^{\frac{1}{2}}$, and let $I = \int_0^4 \sqrt{x}\,dx$. The graph of $f(x)$ is given below.

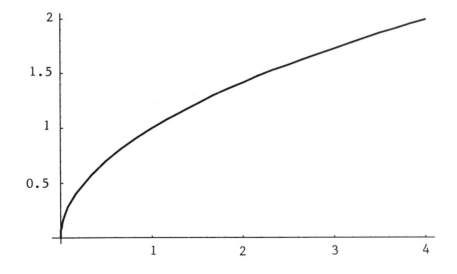

a) Compute I exactly using the Fundamental Theorem of Calculus.

b) Write out the first three and last terms of L_{10} and R_{10}.

$$L_{10} = \underline{\hspace{1cm}} + \underline{\hspace{1.5cm}} + \underline{\hspace{1.5cm}} + \ldots + \underline{\hspace{1.5cm}}$$

$$R_{10} = \underline{\hspace{1cm}} + \underline{\hspace{1.5cm}} + \underline{\hspace{1.5cm}} + \ldots + \underline{\hspace{1.5cm}}$$

c) Write L_{10} using sigma notation.

d) Technology says $L_{10} = 4.88407$ and $R_{10} = 5.68407$. Using only this information, find T_{10}.

e) Fill in the blanks: L_{10} is an underestimate for I and R_{10} is an overestimate for I because $f(x) = \sqrt{x} = x^{1/2}$ is \underline{\hspace{4cm}}.

f) Fill in the blanks: T_{10} is too \underline{\hspace{2cm}} (small or large) because $f(x) = \sqrt{x} = x^{1/2}$ is \underline{\hspace{4cm}}.

g) What happens when you try to find the error bound for L_{10} using the bound involving K_1?

h) Find the actual error for L_{10}.

G,N-2 9. Let $f(x) = (.5)^x$ and $I = \int_0^3 f(x)\,dx$. The graph of $f(x)$ is given below.

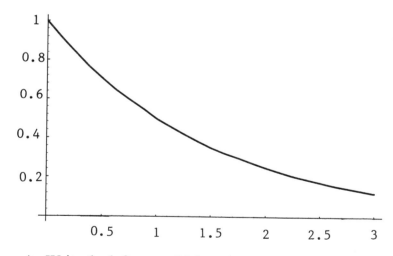

a) Write the left sum which approximates I with 20 subdivisions using sigma notation.

b) Fill in the blanks: The approximation using L_{20} is too \underline{\hspace{3cm}} (small or large) because \underline{\hspace{6cm}}.

c) Calculate both the Theorem 1 error bound (without K_1), and the Theorem 2 error bound (with K_1).

N-1   10. Let $I = \int_{-1}^{2} e^{-x^2} dx$.

a) Compute L_6 by hand. You may use a scientific calculator.

b) Use the bound $\left| I - L_n \right| \leq \dfrac{K_1(b-a)^2}{2n}$ to estimate the error from using L_6.

G,N-1 11. Let $I_1 = \int_0^2 x^2 dx$ and $I_2 = \int_0^2 e^x dx$. The graphs of $f_1(x) = x^2$ and $f_2(x) = e^x$ are given below.

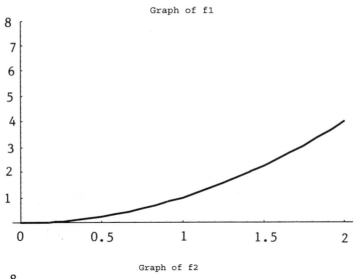

Graph of f1

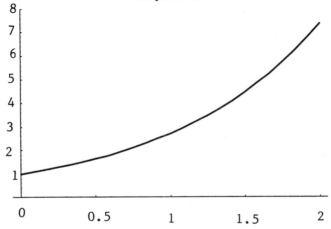

Graph of f2

a) Let E_1 and E_2 be the estimated error bounds for using a left approximating sum with 2 subintervals to approximate I_1 and I_2, respectively. Using only the graph, tell which is larger, E_1 or E_2. Explain your answer.

b) Calculate the actual errors for using a left approximating sum with 2 subintervals to approximate I_1 and I_2. Which is larger?

G,N-1 12. Let $I = \int_0^{10} f(x)\,dx$. The graph of the derivative of f is given below.

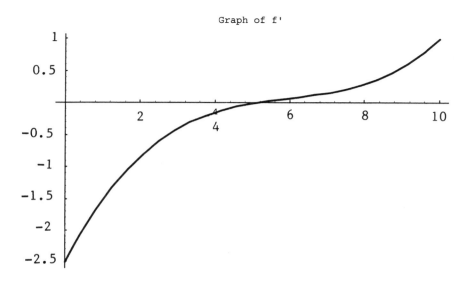

Graph of f'

a) Use the graph to estimate the error in approximating I with R_{10}.

b) Could you use the error estimate for monotone functions to estimate the error in approximating I with R_{10}? Explain.

SECTION 3: TRAPEZOID SUMS, MIDPOINT SUMS, AND THE SECOND DERIVATIVE

N,S-2 13. Consider $f(x) = x^9$ on [0, 1]. Graph f on your calculator, then sketch the graph.

a) Calculate $I = \int_0^1 f(x)\,dx$ exactly.

b) Calculate L_1 and T_1 by hand.

c) Calculate the errors for L_1 and T_1 using your answers to parts (a) and (b). Which approximation has the larger error? Does this contradict what you've learned? Explain.

d) How large does n have to be so that T_n is a better approximation to I than L_n? Show work to support your answer (calculate T and L for such an n, then calculate the errors associated with them).

N,S-2 14. Consider $\int_1^2 \dfrac{1}{x^2}\,dx$.

a) Calculate the integral exactly.

b) Using technology, find T_5.

c) Using your answers from (a) and (b), calculate the actual error from using T_5.

d) Estimate the error for T_5 using the estimate for monotone functions.

e) Estimate the error for T_5 using the estimate which requires K_2 .

N-2   15. Let f be the function graphed below.

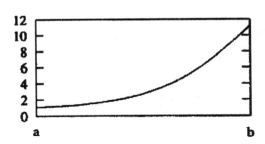

Estimates of $\int_a^b f(x)\,dx$ were computed using the left, right, midpoint, and trapezoid rules. The answers recorded were 9.73601, 9.61105, 9.73677, 9.86249. Which rule produced which estimate? Justify your answer.

N-1   16. The table below gives speedometer readings at various times over a one hour interval.

speed (mph)	40	46	38	0	28
time (min)	0	15	30	45	60

Estimate the total distance traveled using a trapezoid approximating sum with 4 subintervals.

N-2   17. Let $f(x) = 1/x$ and let $I = \displaystyle\int_1^3 \frac{1}{x}\, dx$. The graph of $f(x) = 1/x$ is given below.

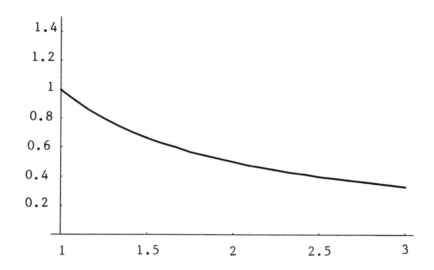

a) Calculate M_2 by hand using only a scientific calculator.

b) Calculate T_2 by hand using only a scientific calculator.

Technology says $L_{100} = 1.105308584$ and we want to know how accurate this approximation is. We have two error bounds.

Theorem 1 says $|I - L_n| \le |f(b) - f(a)| \cdot \dfrac{b-a}{n}$, and

Theorem 2 says $|I - L_n| \le K_1 \cdot \dfrac{(b-a)^2}{2n}$

c) Does Theorem 1 apply here? Explain.

d) Find a value for K_1 in Theorem 2.

e) Calculate both error bounds.

N-1   18. Use the Trapezoidal rule with four subintervals to approximate $\int_0^2 \sin x \, dx$. Do the computations by hand using only a scientific calculator.

SECTION 4: SIMPSON'S RULE

N,S-3   19. Consider the integral $I = \int_0^2 \sin(2x) \, dx$.

a) Use the Fundamental Theorem of Calculus to evaluate the integral exactly.

b) Use technology to estimate the value of the integral using Simpson's Rule with 20 subintervals.

c) Calculate the approximation error using your answers to (a) and (b).

d) Calculate the theoretical error bound using the error bound formula for Simpson's Rule. Compare this theoretical error with the actual error in (c).

e) What is the smallest value of n that makes the Simpson approximation S_n accurate to within .001? ($|I - S_n| \le .001$)

N-2   20. Let $I = \int_1^2 f(x) \, dx$, where some values of the function f are given below:

x	1.00	1.25	1.50	1.75	2.00
$f(x)$	1.23	1.51	1.65	1.48	1.19

a) Calculate estimates of I using L_4, M_2, T_4, S_4.

b) A plot of the data makes it seem reasonable to assume that the graph of f is concave down on the interval $[1, 2]$. Use this assumption and your results from (a) to find upper and lower bounds on I. Justify your choices.

N,S-3   21. Let $I = \int_0^2 e^{x^2} \, dx$. The graph of $f(x) = e^{x^2}$ is given below.

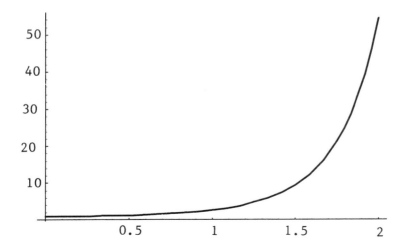

a) Compute M_2, T_2, S_4 by hand using only a scientific calculator.

b) Bound the approximation error made by using S_4. Note: If $f(x) = e^{x^2}$, then
$$f^{(4)}(x) = 4e^{x^2}(3 + 6x + 6x^2 + 4x^4).$$

c) What is the smallest value of n necessary to guarantee that S_n approximates I to within 1×10^{-5}?

G,N-3 22. Let $f(x) = \sin(x^2)$ and $I = \int_0^{\sqrt{\pi}} f(x)\,dx$. If you were going to approximate I using each of the following methods, what value of n would you need in order to guarantee an accuracy of .01? Use the graphs to help answer the questions.

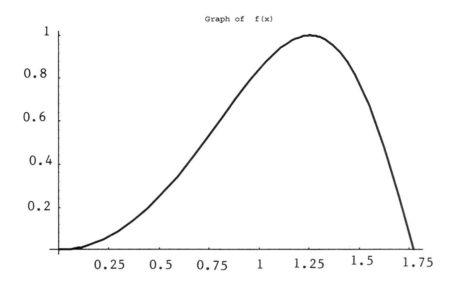

Graph of f(x)

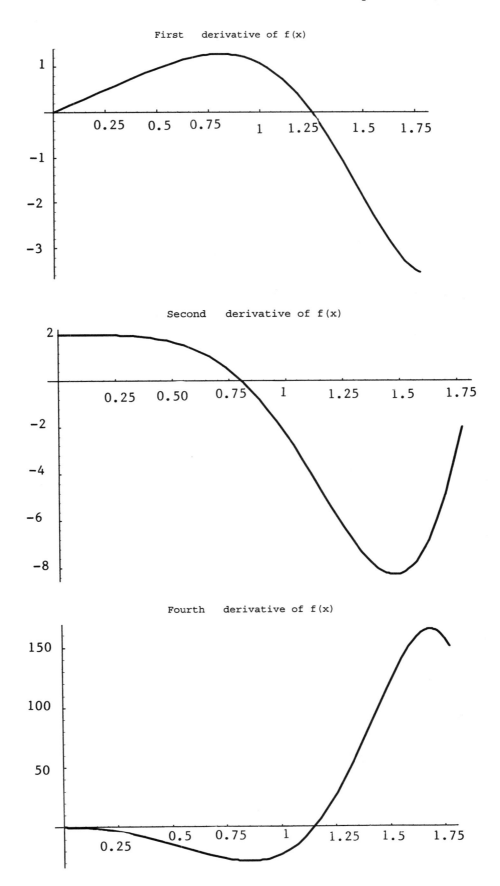

First   derivative of f(x)

Second   derivative of f(x)

Fourth   derivative of f(x)

a) left sum

b) midpoint rule

c) trapezoid rule

d) Simpson's rule

N-2   23. Let $f(x) = 1 - (x-1)^4 = -x^4 + 4x^3 - 6x^2 + 4x$. The graph of f is given below.

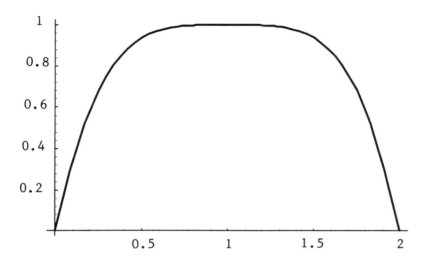

a) Use S_2 (calculated by hand) to approximate $I = \int_0^2 f(x)\,dx$.

b) Find K_4.

c) Find the error bound for approximating I with S_2.

d) Find the actual error.

e) Find the smallest value of n necessary so the Simpson error bound is $\leq .0001$.

N-3   24. Consider the graph of $f(x) = \arccos x = \cos^{-1} x$ given below. We wish to approximate $I = \int_{-1}^1 \arccos x\,dx$.

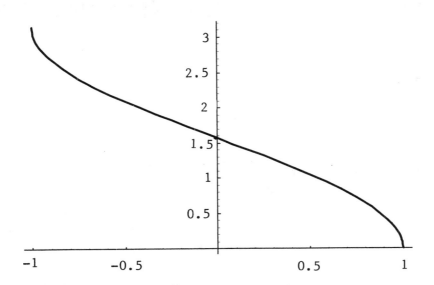

a) Calculate the right sum with 4 rectangles, and draw the rectangles on the graph.

b) Fill in the blank: This approximation is too _____ (small or large). Explain your answer.

c) Calculate the trapezoid and midpoint approximations, by hand, using 4 subintervals.

d) Calculate the Simpson's approximation, by hand, using 4 subintervals.

e) All of the error bound formulas involve derivatives. Explain why there is a problem here - in particular, what is $f'(1)$, and what does this make K_1?

f) Someone claims that both T_4 and M_4 are <u>exact</u>. Do you agree? Explain.

CHAPTER 8: USING THE DEFINITE INTEGRAL

SECTION 1: INTRODUCTION

S-1   1. Consider the integral $I = \int_0^\pi \sqrt{1 + \cos^2 x}\, dx$.

       a) I can be thought of as the area under some curve. Decide which curve, and draw the region in question. Estimate the area under the curve.

       b) I can also be thought of as the length of a graph $y = f(x)$, from $x = 0$ to $x = \pi$. Find and plot such a function f. Estimate the arc length.

       c) Use technology to compute T_{100} (a trapezoid approximating sum with 100 subintervals). This should be a good estimate for I. Compare this with your estimates in a) and b).

S-3   2. Recall the distance formula: The distance between two points (x_1, y_1) and (x_2, y_2) is given by

$$d\big[(x_1, y_1), (x_2, y_2)\big] = \sqrt{(x_2 - x_1)^2 + (y_2 - y_1)^2} \ .$$

       a) Write the equation of the straight line through the two points (x_1, y_1) and (x_2, y_2) in the form $y = f(x)$.

       b) Calculate $f'(x)$.

       c) Using your answer to (b) and the formula for arc length, calculate the arc length of the line segment.

       d) Using algebra, prove that your answer to (c) is the same as the result given by the distance formula.

SECTION 2: FINDING VOLUMES BY INTEGRATION

S-2   3. a) Sketch the region bounded by $y = 2x$, $y = x + 1$ and $x = 0$.

       b) Calculate the volume of the solid of revolution obtained when this region is revolved around the x-axis.

S-3   4. Find the volume of a solid whose base lies along the circle $x^2 + y^2 = 1$ and whose cross-sections perpendicular to the x-axis are squares.

S-2   5. Find the volume which results when the region bounded by $y = 1/x$ and $y = 1/x^2$ and the line $x = 2$ is revolved around the x-axis.

S-1   6. Consider the region shown below.

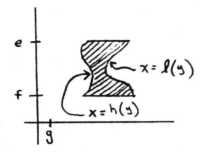

    a) Write an integral expression that can be used to determine the volume of the solid that results when the region is revolved about the y-axis.

    b) Write an integral expression that can be used to determine the volume of the solid that results when the region is revolved about the line $x = g$.

S-3   7. The base of a certain solid is the circle $x^2 + y^2 = 36$ and each cross section perpendicular to the y-axis is an equilateral triangle. Find the volume of the solid.

S-3   8. The base of a solid is an isosceles right triangle whose legs are each 4 units long, as shown below. Any cross section cut perpendicular to the x-axis is a semi-circle. Find the volume of this solid.

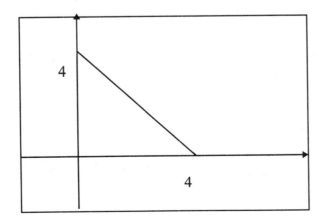

SECTION 3: ARCLENGTH

S-1   9. Set up the integral needed to find the length of the curve $y = \ln x$ from $x = .5$ to $x = 3$. Do not evaluate.

S-2   10. If the curve $y = f(x)$ has the property that $y' = \sqrt{x^3 - 1}$, find the length of the curve from $x = 4$ to $x = 9$.

S-3   11. Find the arclength of the curve $f(x) = \frac{1}{3}(x - 3)\sqrt{x}$ from $x = 0$ to $x = 3$.

S-3   12. Let C be the curve described by $y = \dfrac{x^3}{6} + \dfrac{1}{2x}$.

   a) Sketch the curve C and give an estimate of its length from $x = 2$ to $x = 3$.

   b) Compute the length of C exactly.

S-3   13. Let $F(x) = \displaystyle\int_0^x \sqrt{e^{2t} - 1}\, dt$. Find the arc length of the curve $F(x)$ for $0 \le x \le 2$.

SECTION 4: WORK

S-2   14. Ben's yo-yo weighs 0.2 pounds and has 2 feet of string which weighs 0.01 pounds per foot. How much work does he do yanking the yo-yo all the way up from its maximum extension?

S-2   15. A spring whose natural length is 15 ft. exerts a force of 60 lb. when stretched to a length of 20 ft.

   a) Find the spring constant.

   b) Find the amount of work done in compressing the spring to a point that makes the spring 3 ft. shorter than the natural length.

   c) Find the amount of work done in stretching the spring from a length of 20 ft to a length of 30 ft.

S-2   16. A 50 ft. length of steel chain weighing 25 lb / ft. is dangling from a pulley. How much work is required to wind the chain onto the pulley?

S-3   17. A cylindrical tank of radius 8 ft. and a height of 20 ft. is filled with water. Write an appropriate integral that could be used to determine the amount of work required to pump the water to a point seven feet above the upper rim of the cylinder. Assume that the top of the tank is 0 and down is the positive direction.

S-3   18. Find the natural length of a heavy metal spring, given that the amount of work done in stretching it from a length of 2 ft. to a length of 3 ft. is one-half the work done in stretching it from a length of 3 ft. to a length of 4 ft.

S-3   19. A conical container (vertex down) of radius 4 ft. and height 8 ft. is full of a liquid with density 72 lb. per cubic foot. Find the work done in pumping the top half (by height) of the liquid to a point 5 feet above the tank.

S-3   20. A leaky 5 lb. bucket is lifted from the ground into the air by pulling in 20 ft. of rope at a constant speed. The rope weighs 0.08 lb/ft. The bucket starts with 2 gal. of water (16 lb.) and leaks at a constant rate. It finishes draining just as it reaches the top. How much work is spent raising the bucket, water and rope?

SECTION 5: PRESENT VALUE

S-1   21. Heidi will enter college when she is seventeen years old. Her mother has calculated that she will need $30,000 to pay for the first year of college. She is presently seven years old and has $10,000 invested. What interest rate will be required on her investment to have enough money for the first year of college?

S-2   22. Find the present value of a continuous income stream of $4,000 per month beginning in one year and lasting for 10 months. Assume an annual interest rate of $r = .06$.

S-1   23. The result of a certain "present value" calculation yielded $\int_{2}^{4} 100\sqrt{t}\, e^{-0.07t}\, dt$ dollars.

(The units of t are years.) What is the meaning of the $100\sqrt{t}$, of the 2 and the 4, and the 0.07? Your answer should be in words that could be understood by somebody who knows nothing about calculus.

S-2   24. Scott has $10,000 to invest. He wants to double his money in 5 years. If the inflation rate stays constant at 7% for those 5 years, what nominal interest rate does Scott need to double his money?

CHAPTER 9: MORE ANTIDIFFERENTIATION TECHNIQUES

SECTION 1: INTEGRATION BY PARTS

S-2   1. Evaluate: $\int x \ln 5x \, dx$

S-3   2. Evaluate: $\int \ln 5x \, dx$

S-1   3. Evaluate: $\int x^{99} \ln x \, dx$

S-2   4. Evaluate: $\int x^3 e^x dx$

S-3   5. Evaluate: $\int x^2 \ln \sqrt{x} \, dx$

S-3   6. Evaluate: $\int e^t \sin 5t \, dt$

S-3   7. Evaluate: $\int \frac{\ln x}{x^2} \, dx$

S-1   8. Evaluate: $\int x \sin x \, dx$

SECTION 2: PARTIAL FRACTIONS

S-2   9. Evaluate: $\int \frac{3x - 2}{(x - 1)(x^2 + 1)} \, dx$

S-2   10. Evaluate: $\int \frac{6}{(x - 2)(x^2 + 1)} \, dx$

S-2   11. Evaluate: $\int \frac{2x + 3}{(x + 1)^2} \, dx$

S-3   12. Evaluate: $\displaystyle\int_{2}^{3} \frac{x^2 + x + 1}{x^3 - 1}\, dx$

S-1   13. Evaluate: $\displaystyle\int \frac{dx}{x^2 - 5x + 6}$

S-1   14. Evaluate: $\displaystyle\int \frac{dx}{x(x^2 + 5)}$

S-1   15. Evaluate: $\displaystyle\int \frac{x}{x^2 + 5x + 6}\, dx$

S-2   16. Break $\displaystyle\int \frac{4x^2 - 3x - 4}{x^3 + x^2 - 2x}\, dx$ into the sum of three integrals whose integrands have irreducible denominators.

SECTION 3: TRIGONOMETRIC ANTIDERIVATIVES

S-2   17. Evaluate: $\displaystyle\int \frac{x}{1 + 9x^4}\, dx$

S-3   18. Evaluate: $\displaystyle\int \frac{x}{\sqrt{x^4 - 25}}\, dx$

S-3   19. Evaluate: $\displaystyle\int_{0}^{\ln\sqrt{3}} \frac{e^x}{1 + e^{2x}}\, dx$

S-3   20. Evaluate: $\displaystyle\int \frac{x + 2}{x^2 + 2}\, dx$

S-3   21. Evaluate: $\displaystyle\int \sqrt{25 - x^2}\, dx$

S-3   22. Evaluate: $\displaystyle\int \sin^3(2x)\, dx$

S-3   23. Evaluate: $\displaystyle\int \frac{x^3}{\sqrt{9 - x^2}}\, dx$. Optional Hint: Let $x = 3\sin u$.

CHAPTER 10: IMPROPER INTEGRALS

SECTION 1: WHEN IS AN INTEGRAL IMPROPER?

S-2   1. Determine if the improper integral $\displaystyle\int_2^\infty \frac{\ln x}{x}\,dx$ converges or diverges. If it converges, evaluate it.

S-2   2. Determine if the improper integral $\displaystyle\int_1^\infty \frac{dx}{x\sqrt{1+x^2}}$ converges or diverges. If it converges, evaluate it.

S-1   3. a) Explain why $\displaystyle\int_1^\infty \frac{1}{x^{2/3}}\,dx$ is improper.

        b) Determine if the integral is convergent or divergent. If it is convergent, evaluate it.

S-1   4. a) Explain why $\displaystyle\int_5^\infty \frac{1}{(x-1)^{3/2}}\,dx$ is improper.

        b) Determine if the integral is convergent or divergent. If it is convergent, evaluate it.

S-1   5. a) Explain why $\displaystyle\int_0^1 \frac{1}{x^3}\,dx$ is improper.

        b) Determine if the integral is convergent or divergent. If it is convergent, evaluate it.

S-2   6. Determine if the improper integral $\displaystyle\int_0^\infty \frac{x^2}{4x^3+5}\,dx$ converges or diverges. If it converges, evaluate it.

S-1   7. Determine if the improper integral $\displaystyle\int_0^1 \frac{1}{x^{1/4}}\,dx$ converges or diverges. If it converges, evaluate it.

S-3   8. Use integration by parts to show: $\int_a^\infty e^{-x^2}\,dx = \dfrac{e^{-a^2}}{2a} - \dfrac{1}{2}\int_a^\infty \dfrac{e^{-x^2}}{x^2}\,dx$. Optional Hint:

Choose $u = 1/x$ and $dv = xe^{-x^2}\,dx$.

SECTION 2: DETECTING CONVERGENCE, ESTIMATING LIMITS

S-2   9. Use comparison to determine if $\int_1^\infty \dfrac{1}{\sqrt{1+x^3}}\,dx$ is convergent or divergent.

S-2   10. Use comparison to determine if $\int_2^\infty \dfrac{x}{\sqrt{x^3-2}}\,dx$ is convergent or divergent.

S-3   11. Show that the improper integral $\int_0^\infty \dfrac{e^{-x}}{\sqrt{x}}\,dx$ converges.

S-2   12. Use comparison to determine if $\int_1^\infty \dfrac{4}{x(x+1)}\,dx$ is convergent or divergent.

S-2   13. Use comparison to determine if $\int_0^\infty \dfrac{dx}{x^3+e^{-x}}$ is convergent or divergent.

SECTION 3: IMPROPER INTEGRALS AND PROBABILITY

S-2   14. If a math test is normally distributed with mean 75 and standard deviation 5,

a) set up the integral for determining the probability of getting a score between 62 and 70.

b) change the raw scores to standard z-scores and set up the integral for determining the probability of getting a score between 85 and 87 on the test.

S-3   15. The probability density function for the duration of telephone calls within a certain city is $p(x) = 0.4e^{-0.4x}$, where x denotes the duration in minutes of a randomly selected call.

a) What is the probability that a call will last between 1 and 2 minutes?

b) What is the expected length of a call?

c) Find the distribution function.

S-3   16. Consider a group of people who have received treatment for a disease such as cancer. Let t be the survival time, the number of years a person lives after receiving treatment. The density function giving the distribution of t is $p(t) = Ce^{-Ct}$ for some positive constant C.

a) What is the practical meaning of the probability distribution function

$$P(t) = \int_0^t p(x)\,dx\,?$$

b) The survival function, $S(t)$, is the probability that a randomly selected person survives for at least t years. Find $S(t)$.

c) Suppose a patient has a 70% probability of surviving at least two years. Find C.

SECTION 4: L'HOPITAL'S RULE: COMPARING RATES

S-2   17. Determine if the improper integral $\int_0^1 (-\ln x)\,dx$ converges or diverges. If it converges, evaluate it.

S-2   18. Evaluate the following limit: $\lim_{x\to\infty} \dfrac{1+2x}{\sqrt{x}}$.

S-2   19. Evaluate: $\lim_{x\to\pi/2} \dfrac{1-\sin x}{x - \pi/2}$.

S-3   20. Evaluate: $\lim_{k\to\infty} \sqrt{k^2 + k} - k$.

S-2   21. Evaluate: $\lim_{x\to 0} \dfrac{e^{2x} - 1}{\sin(3x)}$,

S-2   22. Evaluate: $\lim_{x\to 0} \dfrac{3^{\sin x} - 1}{x}$.

S-2   23. Evaluate: $\lim_{x\to\infty} \dfrac{x^2}{e^{2x}}$.

S-2   24. Evaluate: $\displaystyle\lim_{x\to\infty} e^{-x} x^{\frac{1}{2}}$.

CHAPTER 11: INFINITE SERIES

SECTION 1: SEQUENCES AND THEIR LIMITS

S-1   1. Does the sequence $\left\{\dfrac{3n}{2n-1}\right\}_{n=1}^{\infty}$ converge or diverge? Why? If it converges, to what does it converge?

S-1   2. Does the sequence $\left\{\sqrt{\dfrac{7n}{n-4}}\right\}_{n=5}^{\infty}$ converge or diverge? Why? If it converges, to what does it converge?

S-2   3. Consider the sequence $\{a_1, a_2, a_3, \ldots\}$, given by $\{-100, -10, -1, -.1, -.01, \ldots\}$.

        a) Write an expression for a_n.

        b) Use your answer from part (a) to evaluate a_{101}.

        c) Does the sequence converge? Justify your answer.

        d) If it does converge, find the limit.

S-2   4. a) For which real values of k does the sequence $\left\{\dfrac{kn}{1-n}\right\}_{n=2}^{\infty}$ converge?

        b) For the values in part (a), what does the sequence converge to?

S-2   5. a) For which real values of k does the following sequence converge?

$$\left\{\frac{n}{1-kn}\right\}_{n=\text{first positive integer greater than } \frac{1}{k}}^{\infty}$$

        b) For the values in part (a), what does the sequence converge to?

S-2   6. a) For which real values of k does the following sequence converge?

$$\left\{\frac{n}{k-n}\right\}_{n=\text{first positive integer greater than } k}^{\infty}$$

b) For the values in part (a), what does the sequence converge to?

SECTION 2: INFINITE SERIES, CONVERGENCE, AND DIVERGENCE

S-1   7. Does the series $\sum_{n=1}^{\infty} \frac{3n}{2n-1}$ converge or diverge? Why? If it converges, evaluate it.

S-2   8. Does the series $\sum_{n=1}^{\infty} \frac{n^n}{n!}$ converge or diverge? Which test did you use? If it converges, evaluate it.

S-2   9. Does the series $\sum_{n=1}^{\infty} \left[\sin\left(\frac{1}{n}\right) - \sin\left(\frac{1}{n+1}\right) \right]$ converge or diverge? Which test did you use? If it converges, evaluate it.

S-1   10. Does the series $\sum_{n=1}^{\infty} 2\left(\frac{3}{5}\right)^n$ converge or diverge? Which test did you use? If it converges, evaluate it.

S-2   11. A certain series $\sum_{k=1}^{\infty} a_k$ has partial sums $S_n = \sum_{k=1}^{n} a_k = 7 + \frac{5}{n}$.

   a) Evaluate S_{10}.

   b) Evaluate $\sum_{k=1}^{\infty} a_k$.

   c) Evaluate $\lim_{k \to \infty} a_k$.

S-2   12. Decide whether each of the following conclusions **must** be true, **might** be true, or **cannot** be true, given their hypotheses. Justify your answers.

   a) If $\{a_n\}$ is increasing, then $\sum_{n=1}^{\infty} a_n$ diverges.

   b) If $\{a_n\}$ is decreasing, then $\sum_{n=1}^{\infty} a_n$ converges.

c) If $\{a_n\}$ is increasing and positive, then $\sum_{n=1}^{\infty} a_n$ converges.

d) If $a_n \to 0$ as $n \to \infty$, then $\sum_{n=1}^{\infty} a_n$ converges.

S-2   13. Does the series $\sum_{n=2}^{\infty} \left[\dfrac{1}{(n+1)^2} - \dfrac{1}{n^2} \right]$ converge or diverge? Which test did you use? If it converges, evaluate it.

S-1   14. If $1 + \frac{1}{4} + \frac{1}{9} + \frac{1}{16} + \frac{1}{25} + \ldots$ converges to $\pi^2/6$, to what does $2 - \frac{1}{6} + \frac{1}{9} + \frac{1}{16} + \frac{1}{25} + \ldots$ converge?

S-3   15. Let $a_n = \sum_{k=1}^{n} \dfrac{k}{n^2}$. Why is $\lim_{n \to \infty} a_n = \int_0^1 x \, dx$? Optional Hint: Think of a_n as a Riemann sum.

SECTION 3: TESTING FOR CONVERGENCE; ESTIMATING LIMITS

S-2   16. Does the series $\sum_{n=1}^{\infty} n^2 e^{-n^3}$ converge or diverge? Which test did you use?

S-1   17. Use the comparison test to decide if the series $\sum_{k=1}^{\infty} \dfrac{k! \, 7^k}{2^k}$ converges or diverges. Be sure to tell what series you compared it to.

S-1   18. Use the comparison test to decide if the series $\sum_{k=1}^{\infty} \dfrac{\sqrt{k}}{k^3 + 1}$ converges or diverges. Be sure to tell what series you compared it to.

S-1   19. Use the comparison test to decide if the series $\sum_{k=1}^{\infty} \dfrac{\ln k}{k}$ converges or diverges. Be sure to tell what series you compared it to.

S-2   20. Use the integral test to decide if the series $\sum_{k=1}^{\infty} \dfrac{\ln k}{k}$ converges or diverges. Be sure to show the integral used.

S-2   21. Can the ratio test be used to show that $\displaystyle\sum_{k=2}^{\infty} \frac{\ln k}{k}$ converges? Explain.

S-1   22. Can the ratio test be used to show that $\displaystyle\sum_{k=1}^{\infty} \frac{1}{k^2}$ converges? Explain.

SECTION 4: ABSOLUTE CONVERGENCE; ALTERNATING SERIES

N,S-3 23. Approximate the value of $\displaystyle\sum_{n=1}^{\infty} \frac{(-1)^n}{5^n \cdot \ln(n+1)}$ with an error less than .005.

S-1   24. Decide whether each of the following conclusions **must** be true, **might** be true, or **cannot** be true, given their hypotheses. Justify your answers.

   a) If $\displaystyle\sum_{n=1}^{\infty} a_n$ converges, then $\displaystyle\sum_{n=1}^{\infty} |a_n|$ converges.

   b) If $\displaystyle\sum_{n=1}^{\infty} |a_n|$ converges, then $\displaystyle\sum_{n=1}^{\infty} a_n$ converges.

S-3   25. Consider the series $\displaystyle\sum_{n=2}^{\infty} \frac{n}{n^3 + \sin n}$.

   a) Does the series converge or diverge?

   b) What test(s) did you use?

S-2   26. Consider the series $\displaystyle\sum_{k=4}^{\infty} (-1)^k \frac{4^k}{k^4}$.

   a) Does the series converge or diverge?

   b) Which test did you use?

   c) If it converges, does it converge absolutely or conditionally?

S-1   27. Consider the series $\displaystyle\sum_{k=1}^{\infty} \frac{(-1)^k}{\sqrt[3]{k}}$.

   a) Does the series converge or diverge?

b) Which test did you use?

c) If it converges, does it converge absolutely or conditionally?

S-1   28. Consider the series $\displaystyle\sum_{k=1}^{\infty} \frac{(-2)^k}{3^k}$.

a) Does the series converge or diverge?

b) Which test did you use?

c) If it converges, does it converge absolutely or conditionally?

S-2   29. Consider the series $\displaystyle\sum_{k=1}^{\infty} \frac{\sin(k\pi/2)}{k}$.

a) Does the series converge or diverge?

b) Which test did you use?

c) If it converges, does it converge absolutely or conditionally?

S-1   30. Consider the series $\displaystyle\sum_{k=2}^{\infty} \frac{(-1)^k}{k \ln k}$.

a) Does the series converge or diverge?

b) Which test did you use?

c) If it converges, does it converge absolutely or conditionally?

SECTION 5: POWER SERIES

S-2   31. Find the interval of convergence for $\displaystyle\sum_{n=1}^{\infty} \frac{(x+1)^n}{n \cdot 2^n}$.

S-2   32. Find the interval of convergence for $\displaystyle\sum_{n=1}^{\infty} \frac{(x-2)^n}{n \cdot 3^n}$.

S-2   33. Find the interval of convergence for $\displaystyle\sum_{n=1}^{\infty} \frac{x^n}{n^3+2}$.

S-2   34. Find the interval of convergence for $\displaystyle\sum_{n=2}^{\infty} \frac{n!x^n}{n^2-1}$.

S-2   35. Find the interval of convergence for $\displaystyle\sum_{n=1}^{\infty} \frac{x^n}{n^2+5}$.

S-2   36. Find the interval of convergence for $\displaystyle\sum_{n=2}^{\infty} \frac{n!(x-4)^n}{n^2-1}$.

S-3   37. Suppose that the power series $\displaystyle\sum_{k=0}^{\infty} a_k(x-3)^k$ converges when $x=8$ and diverges when $x=-8$. Indicate which of the following statements **must** be true, **might** be true, and **cannot** be true.

   a) The power series converges when $x=-7$.

   b) The power series converges when $x=12$.

   c) The power series diverges when $x=15$.

   d) The power series diverges when $x=-5$.

S-2   38. Find the interval of convergence for $\displaystyle\sum_{k=0}^{\infty} \left(\tfrac{4}{5}\right)^k x^k$.

SECTION 6: POWER SERIES AS FUNCTIONS

S-2   39. Use series to find $\displaystyle\int \frac{\ln(1+x)}{x}\,dx$. Assume $|x|<1$.

   Optional Hint: $\displaystyle\frac{1}{1+x} = 1-x+x^2-x^3+\ldots$

We know that $\dfrac{1}{1+x} = 1 - x + x^2 - x^3 + x^4 - x^5 + \ldots$ for $x \in (-1,1)$ and

$\sin x = x - \dfrac{x^3}{3!} + \dfrac{x^5}{5!} - \dfrac{x^7}{7!} + \ldots$ for $x \in (-\infty, \infty)$. Use these facts to find a power series and the interval of convergence for the functions given in problems 40 - 43.

S-2   40. $\dfrac{x}{1-x}$

S-2   41. $x^2 \cos x$

S-3   42. $\displaystyle\int (\sin x + \cos x)\, dx$

S-3   43. $\displaystyle\int x^2 \sin x \, dx$

S-2   44. Find the elementary function represented by the power series $\displaystyle\sum_{n=0}^{\infty} \dfrac{x^{2n}}{n!}$, by manipulating a more familiar power series.

S-2   45. Use power series to show that $y = e^{x^2}$ is a solution of the differential equation $y' - 2xy = 0$.

SECTION 7: MACLAURIN AND TAYLOR SERIES

N,S-2 46. a) Find the first two nonzero terms of the Taylor series for $f(x) = (\ln x)^2$ about $a = 1$.

b) Use the result from (a) to approximate $\displaystyle\int_{1}^{1.5} (\ln x)^2 \, dx$.

S-3   47. a) Find the Taylor series centered at 0 for $f(x) = \dfrac{1}{1+x}$.

b) Use the answer from part (a) to find the Taylor series for $\dfrac{x^2}{1+x^3}$.

c) Use the answer from part (b) to find the Taylor series for $\ln(1 + x^3)$.

S-1   48. a) Find the Maclaurin series for the function $y = e^{3x}$ about the point $x = 0$.

   b) Find the radius of convergence for the series you found in (a).

S-2   49. Consider the Maclaurin series for the function $f(x) = \dfrac{1}{1 + x^2}$ given below:

$$f(x) = \sum_{k=0}^{\infty} (-1)^k x^{2k}$$

   Using technology, determine the interval of convergence by plotting f and some partial sums of the series.

S-2   50. Consider the Taylor series given below for the function $f(x) = \ln x$ about the point $x = 1$:

$$f(x) = \sum_{k=1}^{\infty} \frac{(-1)^{k+1}}{k} (x - 1)^k .$$

   Using technology, determine the interval of convergence by plotting f and some partial sums of the series.

S-3   51. a) Find the Maclaurin series for the function $\dfrac{\sin \sqrt{x}}{\sqrt{x}}$. Use this to find the Maclaurin

   series for $\displaystyle\int \frac{\sin \sqrt{x}}{\sqrt{x}} dx$.

   b) This integral can also be evaluated explicitly. Show that
   $\displaystyle\int \frac{\sin \sqrt{x}}{\sqrt{x}} dx = -2 \cos \sqrt{x}$. Find the Maclaurin series for $-2 \cos \sqrt{x}$ and compare it to your answer in part (a). Explain any differences that you find.

CHAPTER 12: DIFFERENTIAL EQUATIONS

SECTION 1: DIFFERENTIAL EQUATIONS: THE BASICS

S-1   1. Is $y = e^{-2x} + x$ a solution of the differential equation $y' + 2y = 2x + 1$? Why or why not?

S-1   2. Is $y = \frac{1}{2}e^x + e^{-2x}$ a solution of the differential equation $y' = 2y + e^{-2x}$? Why or why not?

S-1   3. Is $y = x^4 + x^2$ a solution of the differential equation $x\, y' = 2y + 2x^4$? Why or why not?

S-1   4. a) Guess the solution of the initial value problem: $\dfrac{dy}{dt} = 9y, \quad y(0) = 5$.

   b) Verify your answer to (a) by differentiating.

S-1   5. Solve the following initial value problems.

   a) $\dfrac{dy}{dx} = 2x, \quad y(0) = 3$

   b) $\dfrac{dy}{dx} = 2y, \quad y(0) = 3$

SECTION 2: SLOPE FIELDS: SOLVING DE'S GRAPHICALLY

S-1   6. Solve the initial value problem: $\dfrac{dy}{dx} = 3x^2 - 2x - 1, \; y(0) = 10$.

G,S-2 7. a) Draw the slope field for the differential equation $y' = t + 1$ by hand.

   b) On the slope field, draw the solution which satisfies the IVP: $y' = t + 1$, $y(0) = 0$.

   c) Solve the IVP by hand and find a formula for the curve drawn in (b). Does it agree with your picture?

G-2   8. Match the slope fields shown below with the differential equations from the list:
(Note: In all graphs shown below, t is the horizontal axis, and y is the vertical axis.)

a) $y' = t + 1$ d) $y' = \cos t$

b) $y' = 1 - y$ e) $y' = \cos(t + y)$

c) $y' = \cos y$ f) $y' = y^2 - 1$

i)   iv)

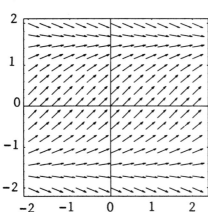   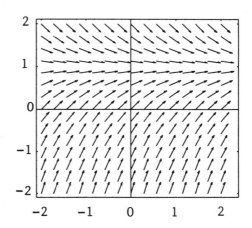

ii)   v)

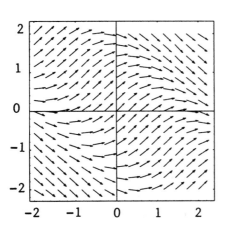   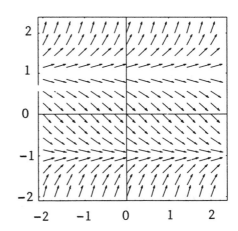

iii)   vi)

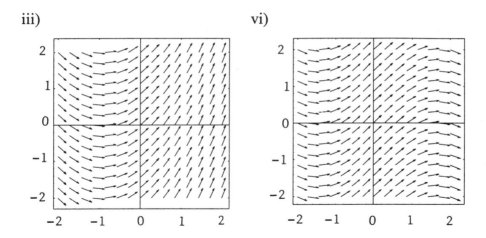

G-2   9.   Below are graphs of several functions of the form $y = 1 - Ce^{-t}$. (These functions are solutions of the DE $y' = 1 - y$).

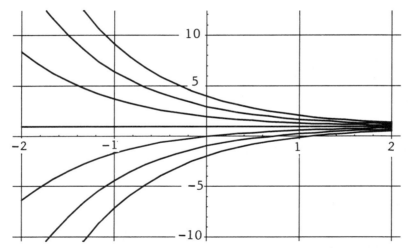

a) One of the curves is a straight line. Which straight line? Label the line with the appropriate value of C.

b) One of the curves passes through the point $(0, 4)$. What value of C corresponds to this curve? How do you know? Label this curve with the appropriate value of C.

c) Estimate (use the graph) the slope at $(0, 4)$ of the curve mentioned in part (b). Does your answer agree with what the DE predicts?

d) Label each curve with the appropriate value of C.

SECTION 3: EULER'S METHOD: SOLVING DE'S NUMERICALLY

N-2   10. Do four steps (step size 0.5) in approximating a solution using Euler's method to the initial value problem $y' = y - t$, $y(0) = 2$.

N-2   11. Consider the IVP $y' = -2y$, $y(0) = 1$.

    a) Guess the exact solution $y(t)$ to the IVP given above. Calculate $y(1)$, $y(2)$, $y(3)$ and $y(4)$.

    b) Use Euler's method, with steps of size 1 to estimate: $y(1)$, $y(2)$, $y(3)$ and $y(4)$. How close are your estimates?

N-2   12. Consider the IVP $y' = -2y$, $y(0) = 0$.

    a) Guess the exact solution $y(t)$ to the IVP given above. Calculate $y(1)$, $y(2)$, $y(3)$ and $y(4)$.

    b) Use Euler's method, with steps of size 1 to estimate: $y(1)$, $y(2)$, $y(3)$ and $y(4)$. What does Euler's method predict about the solution of this IVP? How do the estimates compare with your answers to (a)?

SECTION 4: SEPARATING VARIABLES: SOLVING DE'S SYMBOLICALLY

S-1   13. Solve the differential equation: $\dfrac{dy}{dt} = -0.8(y - 500)$.

S-2   14. A population of honey bees grows logistically. If it is known that a maximum of 1,000 bees can survive in a given hive:

    a) Write a differential equation that represents this situation.

    b) If fifty bees are introduced to this hive when it is new, and there are one hundred bees after two days, find the function that will be used to determine the bee population for any value of t.

S-2   15. a) Verify by direct calculation that $P(t) = C\left(1 + 2e^{-3Ct}\right)^{-1}$ is a solution to the logistic differential equation $P' = 3P(C - P)$.

    b) Find the carrying capacity: $\lim\limits_{t \to \infty} P(t) = $ _____.

c) Calculate $P(0) =$ _____ and $P(1) =$ _____.

d) Use the original differential equation to calculate $P''(t)$. Verify that the inflection point occurs when $P = C/2$.

e) Find t when $P(t) = C/2$.

S-2   16. Solve the initial value problem: $y' = yt$, $y(0) = 1$.

S-2   17. Given the differential equation $y'(t) = k(y - T)$ and the following values:
$T = 50\,°F$, $y(0) = 100\,°F$ and $y(5) = 80\,°F$,

a) Solve the initial value problem

b) Find t when $y(t) = 60\,°F$.

S-2   18. Consider the differential equation: $\dfrac{dy}{dt} = y^2 \sin t$.

a) Show directly that $y = \sec t$ is a solution to this differential equation.

b) Find the solution to the differential equation that satisfies the initial condition $y(0) = 1/2$. Optional Hint: Use the method of separation of variables.

S-3   19. Suppose that at 1:00 one winter afternoon there is a power failure at your house in Minnesota, and your heat does not work without electricity. When the power goes out, it is 68°F in your house. At 10:00 p.m., it is 57°F in the house and you notice that it is 10°F outside.

a) Assuming that the temperature, T, in your home obeys Newton's Law of Cooling, write the differential equation satisfied by T.

b) Solve the differential equation to estimate the temperature in the house when you get up at 7:00 a.m. the next morning. Should you worry about your water pipes freezing?

c) What assumption did you have to make in part (a) about the temperature outside? Given this (probably incorrect) assumption, would you revise your estimate up or down? Why?

S-2   20. Let $P(t)$ represent the number of wolves in a population at time t years, when $t \geq 0$. The population $P(t)$ is increasing at a rate directly proportional to $800 - P(t)$, where the constant of proportionality is k.

a) If $P(0) = 500$, find $P(t)$ in terms of t and k.

b) If $P(2) = 700$, find k.

c) Find the value that $P(t)$ approaches as t gets very large.

CHAPTER 13: POLAR COORDINATES

SECTION 1: POLAR COORDINATES AND POLAR CURVES

S-1   1. Change the equation $y = x^2$ into polar coordinates, writing it in the form $r = f(\theta)$.

S-1   2. Give the rectangular coordinates of the point whose polar coordinates are $\left(4, \frac{\pi}{6}\right)$.

S-1   3. Find a representation for all possible polar coordinates for the rectangular coordinate $(\sqrt{3}, -1)$.

S-2   4. Write the equation $(x - a)^2 + y^2 = a^2$ in polar coordinate notation.

S-3   5. Identify the curve and write the equation $r = 2(1 - \sin\theta)^{-1}$ in rectangular coordinates.

S-2   6. Find the points where the curves $r = 1 - \cos\theta$ and $r = \cos\theta$ intersect.

G-2   7. Sketch the polar curve $r = 2 + 3\cos\theta$. Indicate all lines of symmetry.

G-2   8. Sketch the polar curve $r = 3\cos(6\theta)$. Indicate all lines of symmetry.

G-2   9. Sketch the polar curve $r = 2 - 4\cos\theta$. Label significant points in rectangular coordinates.

G-2   10. Sketch the polar curve $r = 2(1 - \sin\theta)$. Label significant points in rectangular coordinates.

G-2   11. Sketch the polar curve $r = -5\sec\theta$. Label significant points in rectangular coordinates.

G-2   12. Sketch the polar curve $r = -3\sin\theta$. Label significant points in rectangular coordinates.

SECTION 2: CALCULUS IN POLAR COORDINATES

S-3   13. Calculate the area of the region outside the curve $r = 2$, but inside the curve $r = 4\sin\theta$. Be sure to draw the two curves and the region.

S-2   14. Find the area enclosed by the cardioid $r = 1 + \cos\theta$.

G,S-2 15. a) Plot the curve determined by the parametric equations: $x = 2 + 3\cos t$ and
$y = 1 + 3\sin t$ for $0 \le t \le \pi$.

b) Eliminate the parameter t and express the curve in rectangular coordinates.

c) What is the speed of a particle moving along the curve at $t = \pi/2$?

G,S-2 16. Let $r(\theta) = 3\cos(2\theta)$.

a) Sketch a graph of $r(\theta)$.

b) Set up (but do not evaluate) an integral to calculate the area inside one petal of this rose.

S-3   17. Calculate the area inside the outer loop, but outside the inner loop of the polar function $r(\theta) = 2 + 4\cos\theta$.

CHAPTER 14: MULTIVARIABLE CALCULUS: A FIRST LOOK

SECTION 1: THREE-DIMENSIONAL SPACE

G-1   1. Plot the equation $z = y^2 + 1$.

G-2   2. Assume that you are tiny and standing at the point (3, 1, 1), looking towards the
            yz-plane (the orientation of the axes is shown below, and the direction from your feet
            to your head is the positive z-direction). You travel 2 units forward, turn left, and
            travel another 2 units.

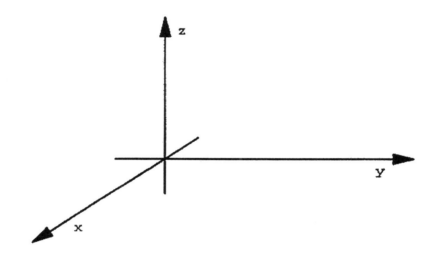

    a) Are you in front of, or behind, the yz-plane?

    b) Are you to the left of, or to the right of, the xz-plane?

    c) Are you above or below the xy-plane?

G,S-1 3. a) Find the equation for a circular cylinder of radius 2 centered along the x-axis.

    b) Sketch the graph of the cylinder.

G,S-1 4. Consider the plane p with equation $4x + 2y + z = 4$.

    a) Sketch the graph of p in the first octant.

    b) Find the trace of p in the xz-plane.

S-2   5. a) Find the distance between the points $(-1, 7, 0)$ and $(3, -2, -3)$.

   b) Use your answer to (a) to find the equation of the sphere which is centered at $(-1, 7, 0)$ and passes through $(3, -2, -3)$.

S-1   6. Suppose the plane p has trace $3x - z = 5$ in the xz-plane and trace $-2y - z = 5$ in the yz-plane. Find the trace of p in the xy-plane.

G-2   7. Plot the equation $z = (x - 2)^2 + 5$ in xyz-space. (Optional Hint: Start in xz-space.)

G-3   8. Plot the equation $x^2 + y^2 + 8x - 6y - 15 = 0$ in xyz-space. (Optional Hint: Start in xy-space and complete the squares.)

SECTION 2: FUNCTIONS OF SEVERAL VARIABLES

G-1   9. Draw at least five level curves of $z = x^2 - 4y^2$, making sure you have all the typical sorts of level curves for this function.

S-2   10. Consider the function $f(x, y) = \sqrt{x^2 - y^2}$.

   a) Find the domain of $f(x, y)$.

   b) Find the range of $f(x, y)$.

S-2   11. Consider the function $f(x, y, z) = |x + y - z|$.

   a) Find the domain of $f(x, y, z)$.

   b) Find the range of $f(x, y, z)$.

G-3   12. Consider the function $f(x, y) = \dfrac{1}{\sqrt{x^2 + y^2}}$.

   a) Find the domain of $f(x, y)$.

   b) Find the range of $f(x, y)$.

   c) Sketch and label the level curves of f corresponding to $z = .2, .3, .4, .5$.

G-2   13. Consider the equation $z = 4x^2 + 9y^2 + 8x + 6y$.

    a) Use technology to plot the graph.

    b) Using the graph from part (a), identify the shape of the level curves.

    c) Find the equation of the level curve corresponding to $z = 1$.

SECTION 3: PARTIAL DERIVATIVES

S-2   14. Find the plane tangent to $z = x^2 + 2y^2$ at the point $(1, -2, 9)$.

N-3   15. The wind-chill factor is a measure of how cold it feels, based on wind conditions and temperature. Shown below is a table of values for the wind-chill factor, $f(w, T)$, as a function of the wind speed w (in miles per hour), and the temperature, T (in degrees Fahrenheit). Use the table to answer the questions below.

T - Temperature (°F)

w, T	35	30	25	20	15	10	5	0
5	33	27	21	16	12	7	0	−5
10	22	16	10	3	−3	−9	−15	−22
15	16	9	2	−5	−11	−18	−25	−31
20	12	4	−3	−10	−17	−24	−31	−39
25	8	1	−7	−15	−22	−29	−36	−44

w - wind (mph)

    a) From the table, you can see that when the temperature is 15°F, and the wind speed is between 5 and 10 mph, the wind chill factor drops by about 3°F for each 1 mph increase in wind speed. Which partial derivative (and at which points) is this statement telling you about?

    b) Estimate $f_w(10, 25)$. What does your answer mean in practical terms?

    c) Estimate $f_T(20, 5)$ and $f_w(20, 5)$.

    d) Estimate a formula for the linear approximation $L(w, T)$ to $f(w, T)$ at $(20, 5)$.

G,S-2 16. Let $h(x, y) = 8xy - \frac{1}{4}(x + y)^4$.

    a) Find $h_x(x, y)$.

b) Find $h_y(x, y)$.

c) If we assume that $y = x$, we get a function of one variable,

$f(x) = h(x, x) = 8xx - \frac{1}{4}(x + x)^4 = 8x^2 - 4x^4$. Sketch rough graphs of $f(x)$, and the function $g(x) = h(x, -x)$, for $-2 \le x \le 2$.

G,N-3 17. Consider the functions numbered 1 to 6, listed below. Each function (except one) is represented by a table, a contour diagram, a graph of the $y = -1$ section ($z = f(x, -1)$), or a 3-D plot. Some functions are represented multiple times. For each of the representations (a) to (l), decide which function numbered 1 to 6 it represents. Which function has no representation?

1. $f(x, y) = x^2 - y^2$

4. $f(x, y) = \dfrac{1}{1 + x^2 + y^2}$

2. $f(x, y) = 6 - 2x + 3y$

5. $f(x, y) = 6 - 2x - 3y$

3. $f(x, y) = \sqrt{1 - x^2 - y^2}$

6. $f(x, y) = \sqrt{x^2 + y^2}$

a)

y↓ x→	−2	−1	0	1	2
−2	2.83	2.24	2.0	2.24	2.83
−1	2.24	1.41	1.0	1.41	2.24
0	2.0	1.0	0.0	1.0	2.0
1	2.24	1.14	1.0	1.41	2.24
2	2.83	2.24	2.0	2.24	2.83

b)

y↓ x→	−2	−1	0	1	2
−2	0.111	0.167	0.2	0.167	0.111
−1	0.167	0.333	0.5	0.333	0.167
0	0.2	0.5	1.0	0.5	0.2
1	0.167	0.333	0.5	0.333	0.167
2	0.111	0.167	0.2	0.167	0.111

c)

y↓ x→	−2	−1	0	1	2
−2	0	−3	−4	−3	0
−1	3	0	−1	0	3
0	4	1	0	1	4
1	3	0	−1	0	3
2	0	−3	−4	−3	0

d)

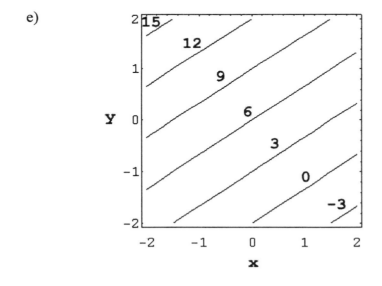

e)

f)

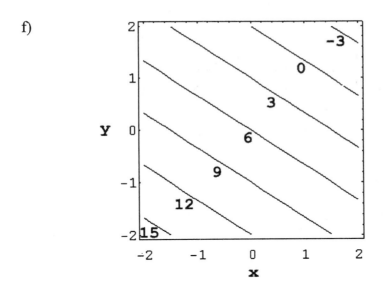

g)

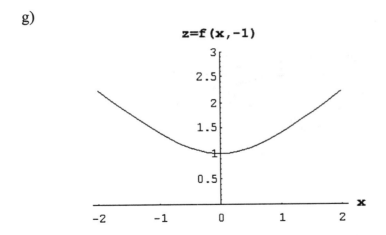

h)

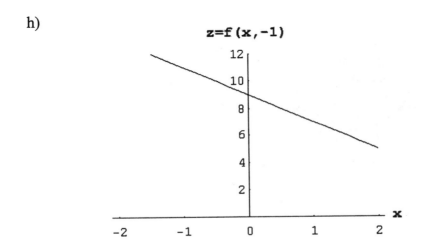

i)

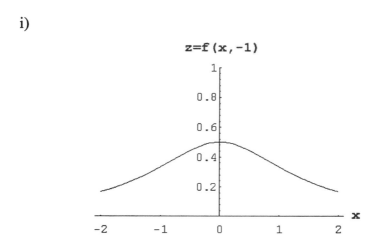

j)

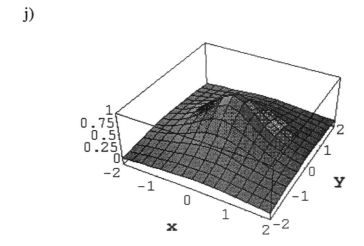

k)

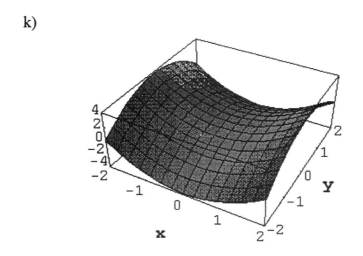

l)

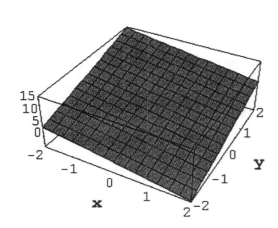

N,S-3   18. For each of the functions (a) to (e) listed below, compute $f_x(1,-1)$ and $f_y(1,-1)$.

a) $f(x, y) = x^2 - y^2$

b) $f(x, y) = 6 - 2x + 3y$

c) $f(x, y) = \dfrac{1}{1 + x^2 + y^2}$

d) $f(x, y) = 6 - 2x - 3y$

e) $f(x, y) = \sqrt{x^2 + y^2}$

f)   Estimate the same two partial derivatives for the function given by the following table:

y↓ x→	−2	−1	0	1	2
−2	2.83	2.24	2.0	2.24	2.83
−1	2.24	1.41	1.0	1.41	2.24
0	2.0	1.0	0.0	1.0	2.0
1	2.24	1.14	1.0	1.41	2.24
2	2.83	2.24	2.0	2.24	2.83

g) Estimate the same two partial derivatives for the function given by the following contour diagram:

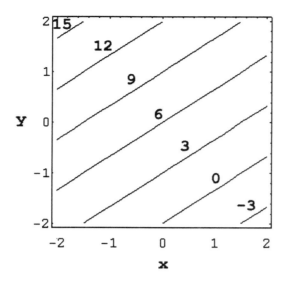

h) Estimate $f_x(1,-1)$ for the function represented by the following $y = -1$ section:

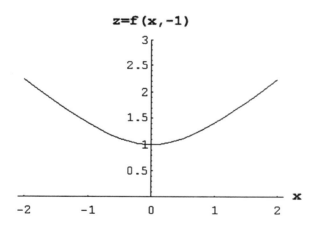

i) Find the linearization $L(x, y)$, which approximates the function represented by the following table, near $(1, -1)$:

y↓ x→	−2	−1	0	1	2
−2	0.111	0.167	0.2	0.167	0.111
−1	0.167	0.333	0.5	0.333	0.167
0	0.2	0.5	1.0	0.5	0.2
1	0.167	0.333	0.5	0.333	0.167
2	0.111	0.167	0.2	0.167	0.111

G,S-2 19. Consider the function $f(x, y) = \sqrt{x^2 + y^2}$

    a) Use technology to make a contour plot with $z = 1, 2, 3, 4$.

    b) Use the plot to approximate $f_x(1, 0)$ and $f_x(1,1)$.

    c) Find $f_x(x, y)$ and $f_y(x, y)$.

    d) Use your answer to part (c) to calculate $f_x(1, 0)$, and $f_x(1, 1)$.

    e) Find the linear approximation to f at the point (1, 1). Explain what the linear approximation is.

SECTION 4: OPTIMIZATION AND PARTIAL DERIVATIVES: A FIRST LOOK

S-2   20. Minimize $f(x, y) = x + 2y$ on the disk $x^2 + y^2 \leq 1$.

S-3   21. Maximize $xy + z$ on the portion of the plane $x + y + z = 1$ which lies in the positive octant (i.e., $x, y, z \geq 0$).

S-2   22. Find all stationary points of $f(x, y) = \frac{1}{x} - \frac{1}{y} + 4x - y$ and identify them as relative maxima, relative minima, or saddle points.

G,N-2 23. Let $h(x, y) = 8xy - \frac{1}{4}(x + y)^4$, for $-2 \leq x \leq 2$ and $-2 \leq y \leq 2$. A contour diagram of $h(x, y)$ is shown below. Light regions correspond to larger z values, and dark regions correspond to smaller (or more negative) z values. Optional Hint: It may be useful to calculate and graph the functions
$$f(x) = h(x, x) = 8xx - \frac{1}{4}(x + x)^4 = 8x^2 - 4x^4, \text{ and } g(x) = h(x, -x).$$

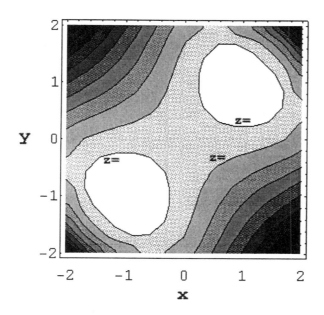

a) On the contour plot, draw in the line $y = x$, and give approximate z values for the three contour lines with a "$z =$" near them.

b) All stationary points of $h(x, y)$ occur along the line $y = x$. Find all such points, exactly, using the partial derivatives of $h(x, y)$ (with the fact above), and decide whether each stationary point is a local maximum, local minimum, or saddle point.

G,S-2 24. Let $g(x, y) = x(x - 2) \cdot \sin y$.

a) Compute $g_x(x, y)$.

b) Compute $g_y(x, y)$.

A contour plot of $g(x, y) = x(x - 2) \cdot \sin y$, for $-0.5 \le x \le 2.5$, and $-3 \le y \le 3$ is shown below. Lighter regions correspond to larger z values.

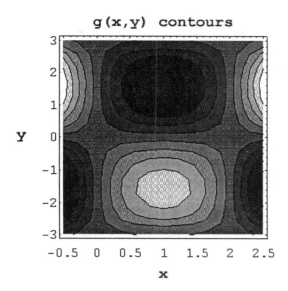

g(x,y) contours

c) There are two stationary points of $g(x, y)$ located along the line $x = 1$. Draw dots on the contour diagram where these two stationary points are located and label each as a local max, local min, or saddle point.

d) There are two stationary points of $g(x, y)$ located along the line $y = 0$. Draw ×'s at these two points and label each as a local max, local min, or saddle point.

e) Using your formulas for g_x and g_y, find the exact coordinates of all four stationary points mentioned above.

S-2   25. You are in charge of erecting a radio telescope on a newly discovered planet. To minimize interference, you want to place it where the magnetic field of the planet is weakest. The planet is spherical with a radius of 6 units. Based on a coordinate system whose origin is at the center of the planet, the strength of the magnetic field is given by $M(x, y, z) = 6x - y^2 + xz + 60$. Where should you locate the radio telescope?

SECTION 5: MULTIPLE INTEGRALS AND APPROXIMATING SUMS

N-2   26. A contour diagram of $f(x, y)$ is shown below. Use it to estimate $\int_{2}^{4} \int_{0}^{1} f(x, y)\, dx\, dy$

with $n = 2$ subdivisions in each direction (4 rectangles total), using the "midpoint" of each subrectangle.

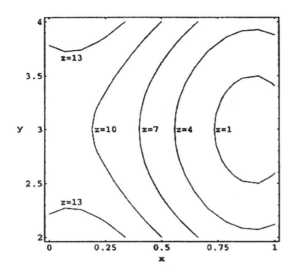

G,N-2 27. Consider the function $f(x, y) = \sqrt{x^2 + y^2}$.

a) Use technology to graph f on the region $[-2, 2] \times [-2, 2]$ and then give a rough estimate of the volume under the curve. Explain how you arrived at your estimate.

b) Use technology to compute the double midpoint sum S_{16} with 16 equal subdivisions (4 in each direction) to approximate this volume.

G,N-2 28. Consider the function $f(x, y) = x y$.

a) Use technology to graph f on the region $[0, 2] \times [0, 2]$ and give a rough estimate of the volume under the curve. Explain how you arrived at your estimate.

b) Compute the double sum with $n = 2$ subdivisions in each direction (4 subdivisions total). Approximate the volume using the point closest to the origin in each subdivision.

c) Give an explanation of why you believe this is larger or smaller than the actual volume.

SECTION 6: CALCULATING INTEGRALS BY ITERATION

S-3   29. Set up and evaluate the volume under $z = xy^2$ over the region bounded by the line $y = 2x$, the x-axis, and the line $x = 3$.

G,S-2 30. Consider the iterated integral $\displaystyle\int_0^6 \int_{x/3}^2 x\sqrt{y^3 + 1}\, dy\, dx$.

   a) Sketch a picture of the region R over which this integral is defined.

   b) There is no elementary antiderivative of $\sqrt{y^3 + 1}$, so switch the order of integration, and evaluate the integral by hand.

S-2   31. Evaluate $\displaystyle\iint_R \sin(y^3)\, dA$, where R is the region bounded by $y = \sqrt{x}$, $y = 2$, and $x = 0$.

SECTION 7: DOUBLE INTEGRALS IN POLAR COORDINATES

S-2   32. Use polar coordinates to evaluate $\displaystyle\int_0^3 \int_0^{\sqrt{9-x^2}} \sin(x^2 + y^2)\, dy\, dx$.

S-1   33. Represent the area inside the curve $r = 4$ and between the lines $\theta = \frac{\pi}{4}$ and $\theta = \frac{\pi}{2}$ using a double integral in polar coordinates. Evaluate the integrals to find the area.

S-2   34. In the first quadrant, find the area between the two curves $r = 1$ and $r = 1 + \cos\theta$.

ANSWERS

CHAPTER 1: FUNCTIONS IN CALCULUS

SECTION 1: FUNCTIONS, CALCULUS-STYLE

1. $f(x) = \begin{cases} -x+1, & -5 \le x < -2 \\ -2 + \sqrt{9 + 8x - x^2}, & -1 \le x \le 7 \end{cases}$

2. $f(-2)$ is undefined, $f(0) = 1$, $f(4) = 3$

3. a) $g(-1) = -1$, $g(0) = 1$, $g(\frac{1}{2}) = 2$, $g(3) = 7$

    b) $g(x) = 2x + 1$

    c)
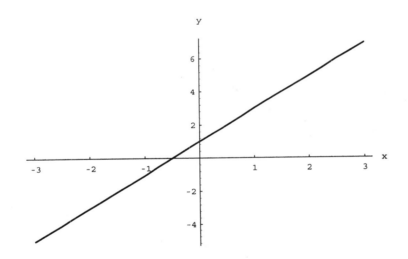

4. a) $A(0) = 4.5$

    b) $A(5) = 17$

    c) $A(3) = 9$

    d) $A(-1) = 5$

    e) $A(x) = \frac{1}{2}x^2 + \frac{9}{2}$

SECTION 2: GRAPHS

5.   The graph of g is the graph of f shifted 5 units to the right and 1 unit upward.

6.   a) 0

      b) 35

      c) [2, 2.25]

      d) [2.25, 2.5]

7.   Translate $y = \ln x$ by $(5 + \ln 3)$ units upward.

8.   a)

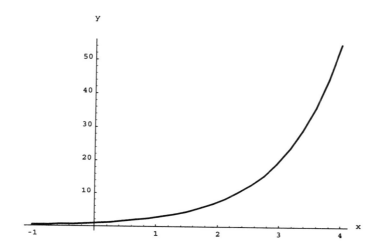

      b)

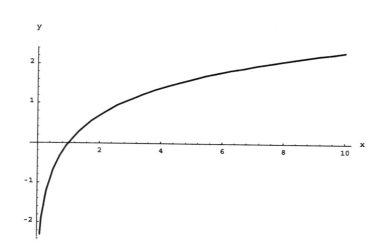

9.

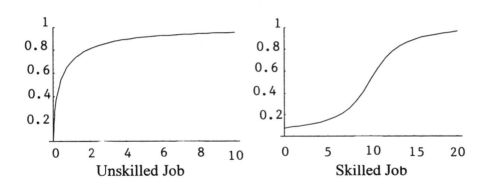

Unskilled Job   Skilled Job

Notice that in the graph for the unskilled job, the worker acquires a large percentage of the required job skills in a very short time, while in the graph for the skilled job, we see that it takes much longer for the worker to attain a high percentage of the necessary skills.

10.   Note: answers may vary.

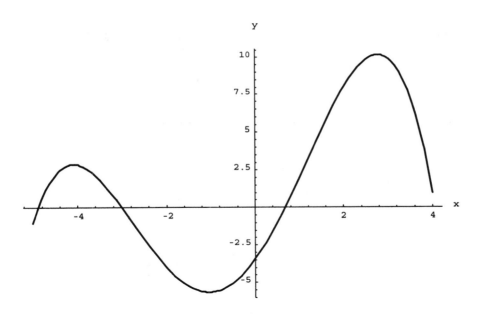

11.   The graph in (b) is the best representation of the cost of living because it is increasing but concave down.

SECTION 3: MACHINE GRAPHICS

12.   One local maximum is at (.86, .56). (Answers may vary). The graph is nearly flat in the final viewing window.

13. Zooming in near (0, 0) until the graph looks like the line $p(x) = x$, we see that it is a good approximation (accurate to within $\pm.01$) on $[-.25, .25]$. On the TI-85, for instance, this can easily be verified by using the "trace" feature.

14. There is only one real root $x \approx 2.37$.

15. Root: -1.8

16. a) Yes. The graph has a local maximum.

    b) The maximum is at approximately $x = 1.0$ and has a value of approximately .5.

SECTION 4: WHAT *IS* A FUNCTION?

17. g is odd, since the graph is symmetric with respect to the origin.

18. g is not periodic, since there is no number p such that $g(x + p) = g(x)$ for all x.

19. Upper bound : 4, Lower bound: -5.

20. Period: π

21. Domain: $[-5, -2) \cup [-1, 7]$; Range: $[-2, 6]$

22. We will show that $(fg)(-x) = -(fg)(x)$.

    $$(fg)(-x) = f(-x) \cdot g(-x) = f(x) \cdot (-g(x)) = -f(x) \cdot g(x) = -(fg)(x)$$

23. f is odd, since the graph is symmetric with respect to the origin.

24. a) 4

    b) 2

    c) 0

    d) True. Since f is periodic with period $\frac{1}{4}$, $f(x - \frac{1}{4}) = f(x)$.

25. Even. The graph is symmetric with respect to the y-axis.

26. Even. The graph is symmetric with respect to the y-axis.

SECTION 5: A FIELD GUIDE TO ELEMENTARY FUNCTIONS

27.   The function is neither even nor odd since the graph is not symmetric with respect to the y-axis or the origin.

28.   The period is $\frac{2\pi}{3} \approx 2.0944$. Note: Answers will vary.

29. $$\frac{3x^2}{4(x+2)(x-5)}$$

30.   a) The function is even since $f(x) = f(-x)$ for all x in the domain.

        b) The function is symmetric about the y-axis since it is even.

        c) The function is not symmetric about the origin since $f(-x) \neq -f(x)$ for every x in the domain. For example, it is not true for $x = 2$.

SECTION 6: NEW FUNCTIONS FROM OLD

31. $h^{-1}(x) = \sqrt[3]{x+4}$

32.   Note: Answers may vary.

        a) Let $f(x) = x^3$, $g(x) = \sin(2x+1)$. Then $h = f \circ g$.

        b) Let $f(x) = x^3$, $g(x) = \sin x$, $p(x) = 2x+1$. Then $h = f \circ g \circ p$.

33. h is even, since $h(-x) = f\big(g(-x)\big) = f\big(g(x)\big) = h(x)$

34.   a) $\ln\left((x+2)^2 + 2\right)$

        b) $\left[\ln(x+2)\right]^2 + 2$

35.   Yes. f^{-1} exists since the graph of f is one-to-one (passes the horizontal line test).

36.   a) $f \circ g$

        b) $g \circ g$

        c) $g \circ f \circ g$

c) $g \circ f \circ g$

d) $g \circ g \circ f$

SECTION 7: MODELING WITH ELEMENTARY FUNCTIONS

37. 2007

38. a) 2.13956

b) 2.13956

c) 2.13956

d) In part (a), we compare average car prices in 1970 and 1960.
In part (b), we compare average car prices in 1995 and 1985.
In part (c), we compare average car prices in 2006 and 1996.

e) They are all the same.

f) Average car price grows by equal ratios over equal time intervals.

39. a) 2626

b) between 7:30 and 8:00

c) 24 hours

CHAPTER 2: THE DERIVATIVE

SECTION 1: AMOUNT FUNCTIONS AND RATE FUNCTIONS: THE IDEA OF THE DERIVATIVE

1. a) 20,800 ft.

    b) 1,920 ft / sec.

    c) 960 ft / sec.

    d) 78,400 ft.

    e) 140 sec.

2. a) 42

    b) −7

    c) 42

    d) 7

3. a) The car is 50 miles north of Chattanooga headed south at 60 mph.

    b) Yes, it is possible. The car could be 100 miles north of Chattanooga headed north at 20 mph.

    c) $V(2) = 0$

    d) $D(t)$ is an amount function and $V(t)$ is its associated rate function. $V(t) = D'(t)$. $V(t)$ tells the instantaneous rate of change of $D(t)$.

    e) The rate of change of velocity (acceleration) is 0. At that instant, the car is neither accelerating nor decelerating.

4. a) 1.5 seconds

    b) 136 ft.

5. a) The gas mileage is decreasing at 55 mph.

    b) Using a linear approximation, 29.55 mpg.

6. $7 \le f(5) \le 17$

7. a) By the Racetrack Principle, $f(2) \le f(0) + 2(4) = 8$ and $f(2) \ge f(0) + 2(1) = 2$.

   By the Racetrack Principle, $f(5) \le f(0) + 5(4) = 20$ and $f(5) \ge f(0) + 5(1) = 5$.

   b) No. By the Racetrack Principle, $f(5) \ge f(2) + 3(1) = 9$.

8. Using the Racetrack Principle and the facts $f'(0) = 7$ and $f''(x) \ge -3$, we see that $f'(x) \ge 7 + x(-3) \ge 1$ for x in [0, 2].

   Using the Racetrack Principle and the facts $f(0) = 0$ and $f'(2) \ge 1$, we see that $f(2) \ge 0 + 2(1) = 2$.

9. Lower Bound: -8

SECTION 2: ESTIMATING DERIVATIVES: A CLOSER LOOK

10. $g'\left(\frac{\pi}{2}\right) \approx -1.57$. The graph looks like a straight line.

11. $f'(6) \approx -1.25$

12. a) .5

    b) -1

    c) The function is not differentiable at 4 since the graph has a sharp corner.

    d)

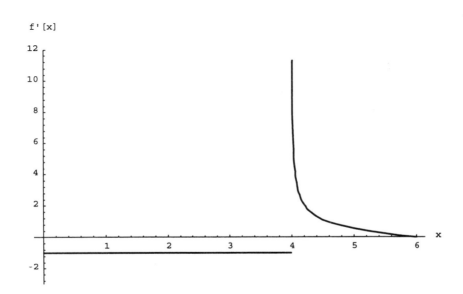

13. a) The missing entries are: $f(-1) \approx 2.5$, $f'(-3) \approx 4$, $f'(0) = -2$

   b) rise: 4, run: 1

   c) rise: -6, run: 3

14. a) The missing entries are:

   $f(0) \approx 0$, $f(1.5) \approx -2.2$, $f(3.0) \approx 0.3$, $f'(0.5) \approx 0.2$, $f'(2.0) \approx -3.2$

   Note: Answers may vary.

   b)

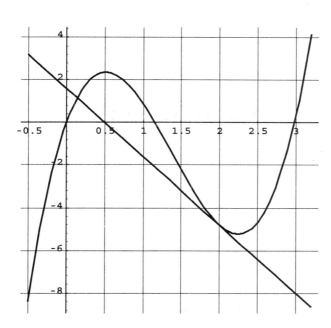

   c) $y = -3.2x + 1.6$ Note: Answers may vary.

SECTION 3: THE GEOMETRY OF DERIVATIVES

15. a) $(-0.5, 0.51)$, $(2.25, 3.2)$

   b) $(0.51, 2.25)$

   c) $(1.38, 3.2)$

   d) $(-0.5, 1.38)$

   e) 1.38

f) −6.6

g) 10

h) −6.4

16.   a) $(-\infty, -0.56), (0.36, 1)$

b) $(-0.56, 0.36), (1, \infty)$

c) $(-0.19, 0.72)$

d) $(-\infty, -0.19), (0.72, \infty)$

e) −0.56 and 1

f) 0.36

g) $y = -x + 2$

17.

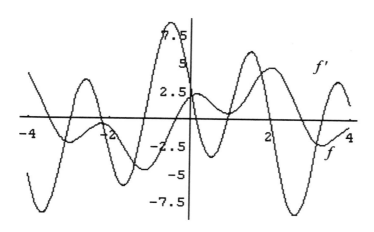

f' is 0 when f has a relative maximum or minimum.

18.

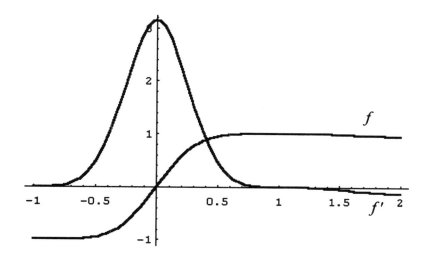

f' is 0 when f has a relative maximum or minimum. Also, f' is positive when f is increasing and f' is negative when f is decreasing.

19. a) $(-\infty, 1)$, $(4, \infty)$

b) Local maximum at 1, local minimum at 4

c) $-\frac{8}{17}$

SECTION 4: THE GEOMETRY OF HIGHER-ORDER DERIVATIVES

20. $(0, 4.67)$

21. b) II only

22. a) $x = 0$, $x = -10$, $x = 10$

b) Relative Maximum: $(0, 0)$; Relative Minimum: $(7.07, -2500)$ and $(-7.07, -2500)$.

c) $(-4.08, -1387.54)$, $(4.08, -1387.54)$

d) Increasing: $(-7.07, 0)$, $(7.07, \infty)$; Decreasing: $(-\infty, -7.07)$, $(0, 7.07)$.

e) Concave Up: $(-\infty, -4.08)$, $(4.08, \infty)$; Concave Down: $(-4.08, 4.08)$.

23. a) $(-1.17, 0.52)$, $(2.90, \infty)$

b) $(-\infty, -1.17)$, $(0.52, 2.90)$

c) $(-\infty, -0.43)$, $(1.93, \infty)$

d) $(-0.43, 1.93)$

e) $(-\infty, -0.43)$, $(1.93, \infty)$

f) $(-0.43, 1.93)$

24.   The dashed function is f and the solid function is f'.

25.   a) 0, 2, and 5

       b) $(-\infty, 0)$, $(2, 5)$

       c) $(0, 2)$, $(5, \infty)$

26.

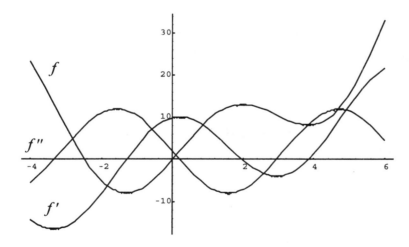

The functions were chosen in this way so that the graph of f' is positive when the graph of f is increasing and negative when the graph of f is decreasing. Also, the graph of f'' is positive when the graph of f is concave up and negative when the graph of f is concave down.

27.   a) $x \approx 5.5$

       b) $x \approx 9$

       c) $(1, 3.5)$, $(7, 10)$

d) $(0, 1)$, $(3.5, 7)$

e) $y = -x + 8$

f) 1.5

g) $(5.5, 9)$

h) $(1, 3.5)$, $(7, 10)$

SECTION 5: AVERAGE AND INSTANTANEOUS RATES: DEFINING THE DERIVATIVE

28. -1

29.   a) $g'(x) = \lim_{h \to 0} \dfrac{g(x + h) - g(x)}{h} = \lim_{h \to 0} \dfrac{\left[f(x + h) + k\right] - \left[f(x) + k\right]}{h}$

$= \lim_{h \to 0} \dfrac{f(x + h) - f(x)}{h} = f'(x)$

b) The graph of g is just the graph of f shifted up or down. The slopes are exactly the same.

30.   a) 28 cm/min

b) Using 2.1, average velocity is 12.61 cm/min, using 2.01, average velocity is 12.0601 cm/min, using 2.001, average velocity is 12.00601 cm/min.

The instantaneous velocity at $t = 2$ is 12 cm/min.

31.   The average velocity over the 6 second time interval is 50 ft/sec. Thus, the driver must have exceeded the 30 mph speed limit.

32.   a) 18 m/sec

b) 12+3h m/sec

c) 12 m/sec

SECTION 6: LIMITS AND CONTINUITY

33.   a) -3

b) 3

34. -1

35. h is not continuous at $x = 2$, because $\lim\limits_{x \to 2^-} h(x) \neq \lim\limits_{x \to 2^+} h(x)$.

36. a) .5

   b) .5

   c) .5

   d) .5

   e) .5

   f) 1

   g) does not exist

   h) .5

   I) $(-\infty, -2) \cup (-2, -1) \cup (-1, 1) \cup (1, \infty)$

37. 4

38. does not exist

39. a) $\left|(3x + 1) - 7\right| = \left|3x - 6\right| = \left|3(x - 2)\right| = 3\left|x - 2\right| < 3\delta = 3 \cdot \frac{\varepsilon}{3} = \varepsilon$

   b) $f = 3x + 1$, $a = 2$, $L = 7$

SECTION 7: LIMITS INVOLVING INFINITY; NEW LIMITS FROM OLD

40. $f'(3) = 1$

41. Can't be done. f can't be both increasing and concave down on $(-\infty, -1)$ without crossing the vertical asymptote.

42. $-\frac{1}{4}$

43. D.N.E.

44. $\lim\limits_{x \to -\infty} \dfrac{e^x + 2}{e^{2x} + 1} = 2$

45.

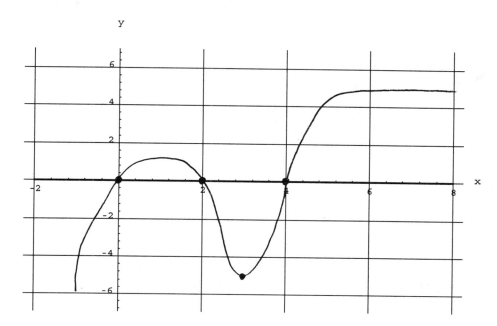

46. 0

47. a) Choose any A.

b) Choose $A = -20$.

c) Choose any A.

CHAPTER 3: DERIVATIVES OF ELEMENTARY FUNCTIONS

SECTION 1: DERIVATIVES OF POWER FUNCTIONS AND POLYNOMIALS

1.   Slope $= 2$

2.   Yes. F is concave down at $x = 6$.

3.   No, since $F(10) - F(8) = \int_{8}^{10} f(x)dx < 7$.

4. $F(x) = x + C$

5. $f(x) = x^4 - 3x^2 + 4$

6.   a) The graph of g is just the graph of f shifted by 4 units to the right. The slope of g at x is the same as the slope of f at $x - 4$. See pictures below.

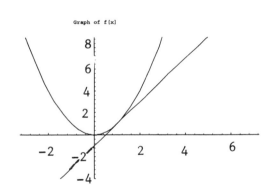
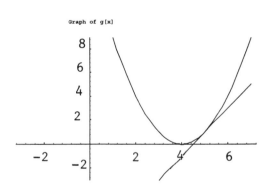

   b) $g'(x) = \dfrac{-3}{(x-4)^4}$

SECTION 2: USING DERIVATIVE AND ANTIDERIVATIVE FORMULAS

7.   NO! You do not stop in time!

8.   Dimensions: $2\,\mathrm{m} \times 2\,\mathrm{m} \times 1\,\mathrm{m}$

9.   Dimensions: radius $= \sqrt[3]{\frac{25}{\pi}} \approx 2$ cm, height $= \dfrac{50}{25^{\frac{2}{3}} \pi^{\frac{1}{3}}} \approx 4$ cm , Note: $h = 2r$

10.   Dimensions: radius $= 3$ in, height $= 12$ in

11.   Dimensions: 8 ft $\times$ 8 ft $\times$ 10 ft

12.   Dimensions: radius $= \sqrt[3]{9} \approx 2.08$ units, height $= 0$ units

13.   a) Dimensions: 3 in $\times \dfrac{\sqrt{91}}{2}$ in

       b) Dimensions: 5 in $\times \dfrac{5\sqrt{3}}{2}$ in

14. $r = \dfrac{2}{\sqrt{\pi}} \approx 1.128$ in

SECTION 3: DERIVATIVES OF EXPONENTIAL AND LOGARITHM FUNCTIONS

15. $f'(x) = \dfrac{1}{x \ln 3}$

16. $k = \frac{1}{3} \ln \frac{5}{2} \approx .3054$

17. $a = \frac{1}{2}, \; b = -\frac{3}{2}$

18. $\dfrac{3}{x} - \dfrac{1}{x \ln 5}$

19. $\dfrac{1}{3} \ln |x^3 + 1| + C$

SECTION 4: DERIVATIVES OF TRIGONOMETRIC FUNCTIONS

20. $\dfrac{-6t^2 e^t \sin(2t^3) - 24t^2 \sin(2t^3) - e^t \cos(2t^3)}{(e^t + 4)^2}$

21.   a) $f'(x) = e^x + \ln 5 \cdot 5^x + \ln \pi \cdot \pi^x - \frac{1}{x} + \sin x$

       b) $f''(x) = e^x + \left(\ln 5\right)^2 \cdot 5^x + \left(\ln \pi\right)^2 \cdot \pi^x + \frac{1}{x^2} + \cos x$

       c) $F(x) = e^x + \dfrac{5^x}{\ln 5} + \dfrac{\pi^x}{\ln \pi} + e^{\pi} \cdot x + x - x \ln x - \sin x + C$

22. $g(x) = \sin x - \dfrac{10^x}{\ln 10} + (\ln 3)x - \dfrac{\pi}{3}x^3 + C$

23.   Derivative $= 2e^x + 2^x \ln 2 + \cos x + 3 \sin x$

       Antiderivative $= e\,x + 2e^x + \dfrac{2^x}{\ln 2} - \cos x - 3 \sin x + C$

24.   a) $\dfrac{4^x}{\ln 4} - \frac{1}{3}\cos(3x) + \frac{3}{2}x^{\frac{2}{3}} + \pi x + C$

       b) $\frac{2}{3}x^{\frac{3}{2}} + \frac{1}{5}e^{5x} + \frac{1}{8}\sin(8x) + e^2 x - \frac{3}{x} + C$

SECTION 5: NEW DERIVATIVES FROM OLD: THE PRODUCT AND QUOTIENT RULES

25.   a) $f'(x) = \dfrac{1 + \cos x + x \sin x}{\cos^2 x}$

       b) $y = 2x$

26. $g(x) = x^2 \sin x + C$

27.   a) -5

       b) 14

       c) $-\frac{3}{4}$

       d) $\frac{26}{25}$

28. f attains a maximum at $x = 3\pi/4$. The maximum value is $f\left(\frac{3\pi}{4}\right) = e^{\frac{3\pi}{4}} \cdot \frac{\sqrt{2}}{2} \approx 7.46$

We are guaranteed the existence of a maximum since every continuous function on a closed bounded interval attains its maximum.

29. a) $p'(-1) = -1$

b) $q'(3) = -\dfrac{1}{2}$

SECTION 6: NEW DERIVATIVES FROM OLD: THE CHAIN RULE

30. $f'(x) = \dfrac{2x\cos(x^2)}{\sin(x^2)} = 2x\cot(x^2)$

31. $F(x) = \frac{1}{3}\sin(3x) - 3\cos\left(\frac{x}{3}\right) + C$

32. a) $g'(-1) = -24$

b) h is decreasing since $h'(2) = -8$.

33. $F(x) = -\frac{1}{2x} + \ln|x| - 2\cos x + \frac{1}{3}e^{3x} + e^{\pi}x + C$

34. a) $f(x) = \dfrac{(x-2)^{54}}{54} + C$, $f''(x) = 53(x-2)^{52}$

b) $f(x) = -9(x+4)^{-\frac{2}{3}} + C$, $f''(x) = -10(x+4)^{-\frac{8}{3}}$

c) $f(x) = -3\cos(x+2) + C$, $f''(x) = 3\cos(x+2)$

d) $f(x) = \frac{1}{3}e^x + C$, $f''(x) = 3e^{3x}$

e) $f''(x) = \frac{1}{x}$

35. $\dfrac{dy}{dx} = 2\sin x \cos x$

36. $f'(x) = x^2 e^{x-2}\left[x\cos(x+7) + (x+3)\sin(x+7)\right]$

129

37.   a) $e^{3x}\left[-4\sin(4x)+3\cos(4x)\right]$

b) $\dfrac{14\left(x^{\frac{5}{3}}-x^4\right)\left[\csc(2x)\cot(2x)\right]+7\left(\frac{5}{3}x^{\frac{2}{3}}-4x^3\right)\csc(2x)}{\left(x^{\frac{5}{3}}-x^4\right)^2}$

c) $2\sin(x\cos x)\cos(x\cos x)(-x\sin x+\cos x)$

38.   Dimensions: $\dfrac{4}{\sqrt{5}}\times\dfrac{1}{\sqrt{5}}$

39.   a) $x>-\frac{1}{2}$

b)   all real numbers

40. $c'(2)=0$

SECTION 7: IMPLICIT DIFFERENTIATION

41. $y=-\frac{2}{5}x-\frac{3}{5}$

42. $\dfrac{dy}{dx}=2$

43.   The point $(3, 1)$ is on the graph because substituting 3 for x and 1 for y satisfies the equation. Slope $=-2/3$.

44. $\dfrac{dy}{dx}=-\dfrac{2x^2+y}{x+y^2}$

45. $y=\frac{1}{2}x+\frac{1}{2}$

46.   slope $=-\frac{4}{3}$

SECTION 8: INVERSE TRIGONOMETRIC FUNCTIONS AND THEIR DERIVATIVES

47. $f'(x) = \dfrac{1}{2x\sqrt{x-1}}$

48. $F(x) = \frac{1}{4}\left(\arctan x\right)^4 + C$

49. $f'\left(\frac{\pi}{12}\right) = \dfrac{6\arctan\left(\frac{\pi}{4}\right)}{1 + \frac{\pi^2}{16}} \approx 2.471$

50. $\arcsin(\tan v) + C$

51. $\dfrac{dy}{dx} = \dfrac{e^x}{\sqrt{1-x^2}} + e^x \arcsin x$

CHAPTER 4: APPLICATIONS OF THE DERIVATIVE

SECTION 1: DIFFERENTIAL EQUATIONS AND THEIR SOLUTIONS

1. No. Because y is not differentiable at $x = -\frac{3}{4}$.

2. a) 10 min 23 sec

   b) 63.6° C.

   c) $-1.48\,°\text{C}\,/\,\text{min}$

3. The death occurred at approximately 4:51 am.

4. k must be negative. y' must be negative since the coffee is cooling and $y - T$ is positive since the coffee has a higher temperature than the room. Therefore, k must be negative so the two sides of the differential equation have the same sign.

5. We must show that v satisfies the differential equation and the initial condition. It is easily shown that $v'(t) = -\dfrac{k}{m}\left(v_0 + \dfrac{mg}{k}\right)e^{-kt/m}$. Consider the left hand side (LHS) and right hand side (RHS) of the differential equation.

   LHS: $mv' = -k\left(v_0 + \dfrac{mg}{k}\right)e^{-kt/m}$

   RHS: $-mg - kv = -mg + mg - k\left(v_0 + \dfrac{mg}{k}\right)e^{-kt/m}$

   $\qquad\qquad = -k\left(v_0 + \dfrac{mg}{k}\right)e^{-kt/m}$

   Thus, LHS = RHS and v satisfies the differential equation.

   Also, $v(0) = -\dfrac{mg}{k} + v_0 + \dfrac{mg}{k} = v_0$ so v satisfies the initial condition.

   Therefore, v is a solution to the initial value problem.

SECTION 2: MORE DIFFERENTIAL EQUATIONS: MODELING GROWTH

6. $706,909.36

7.   a) $P'(t) = .05P(t) - dP(t) = (.05 - d)P(t)$

        b) $P'(t) = (.05 - d)P(t) + 1,000,000$

8.   a)   Since the mixture is being vented at 2,000 cubic meters per minute and the building holds 100,000 cubic meters, we get a rate of $-.02$. Since the building initially has a 1% mixture of natural gas and air, the initial volume is 1,000 cubic meters.

        b) $v(t) = 1,000\, e^{-.02t}$

        c)   301.2 cubic meters

9.   a)   3.87 years

        b)   3.87 years

        c)   3.87 years

        d)   The answers are all the same.

        e)   Doubling time does not depend on the initial amount.

10.   a) $107,946.25

        b) $110,982.01

        c) $111,267.29

        d) $111,276.64

        e) $111,277.04

        f) $111,277.05

        g)   As the time intervals get shorter (*n* increases) in (a) - (e), the answers get closer and closer to the answer in (f).

SECTION 3: LINEAR AND QUADRATIC APPROXIMATION; TAYLOR POLYNOMIALS

11.   a) $f(.25) \approx 1.25$

        b) $f(.25) \approx 1.28125$

12. $f(x) \approx 2x - \frac{4}{3}x^3$. This approximation is accurate on $[-.5, .5]$. Answers may vary.

13. $p(x) = -3x + 4$

14. $p(x) = 1 + \frac{1}{2}(x-1) - \frac{1}{8}(x-1)^2 + \frac{1}{16}(x-1)^3 - \frac{5}{128}(x-1)^4 + \frac{7}{256}(x-1)^5$

15. $f(x) = e^{2x}, \ x_0 = 1$

16.   a) $p(x) = \frac{\pi}{2} + \left(x - \frac{\pi}{2}\right) - \frac{\pi}{4}\left(x - \frac{\pi}{2}\right)^2 - \frac{1}{2}\left(x - \frac{\pi}{2}\right)^3$

        b)   As $x \to \infty$, $p(x) \to -\infty$ and f oscillates between the lines $y = x$ and $y = -x$ increasing without bound.

17.   a) $P_1(x) = 1 + 2x$

        b)

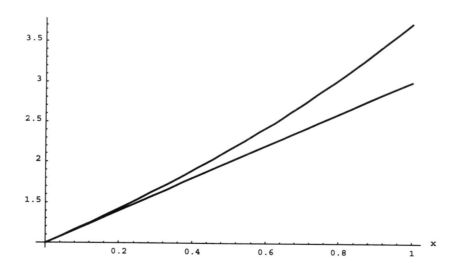

The graph of $f(x)$ is above the graph of $P_1(x)$.

Consider $f(x) - P_1(x) = x + e^x - (1 + 2x) = \left(e^x - x\right) - 1 > 0$ for x in $(0, 1)$.

c) $P_2(x) = 1 + 2x + \frac{1}{2}x^2$

d) $P_2(x) = (1 + e) + (1 + e)(x - 1) + \frac{e}{2}(x - 1)^2$

18. $\quad P_2(x) = x + x^2$

SECTION 4: NEWTON'S METHOD: FINDING ROOTS

19. a)

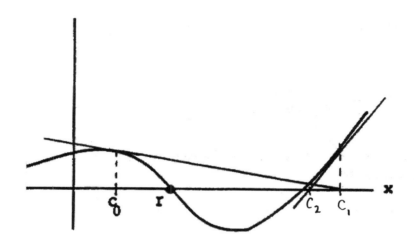

b) If we continue the process, we will find a different root.

20. a) $x_1 = 1.7854, \quad x_2 = 1.5674, \quad x_3 = 1.5708$

The values are approaching $\pi/2$.

b)

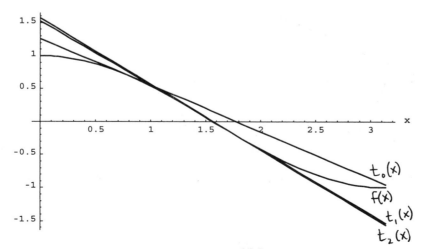

21. In general, $x_{n+1} = x_n - \dfrac{f(x_n)}{f'(x_n)}$. For our function f, $x_{n+1} = x_n - \dfrac{x_n^3 - 3x_n + 1}{3x_n^2 - 3}$. The largest root is approximately 1.532089.

22. a)

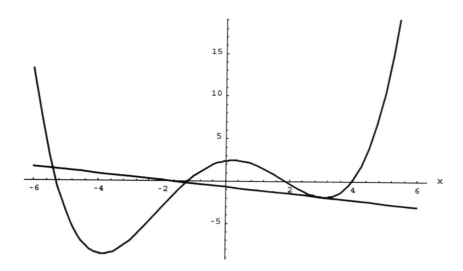

b)

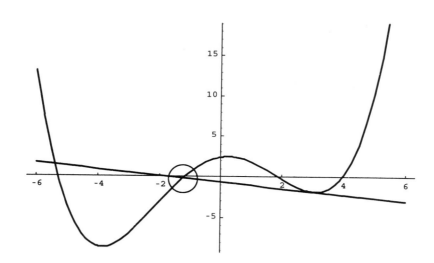

c) The tangent line is nearly horizontal (has a slope close to 0).

23. Solving the equation $e^{-x} = 10 \ln x$ is the same as finding roots of the function $f(x) = e^{-x} - 10 \ln x$. We can use Newton's method to calculate $x_1 = 1.0354826$.

24. 1.16556119

SECTION 5: SPLINES: CONNECTING THE DOTS

25. $S(x) = \begin{cases} -7x^2 + 2x + 2, & 0 \le x \le 1 \\ 15(x-1)^2 - 12(x-1) - 3, & 1 \le x \le 2 \end{cases}$

26. $S(x) = \begin{cases} -2(x-1)^3 + (x-1) + 3, & 1 \le x \le 2 \\ 10(x-2)^3 - 6(x-2)^2 - 5(x-2) + 2, & 2 \le x \le 3 \end{cases}$

27. $a = 1$, $b = -1$, $c = 5$

SECTION 6: OPTIMIZATION

28. The crate has dimensions 2m × 2m × 1m.

29. Maximum Volume: 8π; Height: 6; Radius: 2

30. Maximum Area: $\frac{20}{3} \cdot \sqrt{\frac{5}{3}} \approx 8.61$

31. $\frac{8}{\sqrt{21}} \approx 1.75$ miles below point B

32. To minimize cost, 10 presses should be used.

SECTION 7: CALCULUS FOR MONEY: DERIVATIVES IN ECONOMICS

33. a) $C'(x) = 2$

b) $R'(x) = 10 - \frac{x}{500}$

c) $P(x) = -\frac{x^2}{1,000} + 8x - 7,000$

d) $P'(x) = 8 - \frac{x}{500}$

e) $x = 1,000$ units, and $x = 7,000$ units

f) $x = 4,000$ units

g) Maximum Profit: $P(4,000) = \$9,000$

h) Price: $\dfrac{R(4,000)}{4,000} = \dfrac{\$24,000}{4,000} = \$6$

34.   a) $70

       b) $20,000

35.   a) $64

       b) $28,768

       c) 488

36. $60.72

37. $6.25

SECTION 8: RELATED RATES

38. $\dfrac{dx}{dt} = 1$

39.   You are falling at a rate of 377 ft / min.

40.   10 ft / min.

41.   .24 rad / sec

42.   .063246 ft / min

SECTION 9: PARAMETRIC EQUATIONS, PARAMETRIC CURVES

43.   a) $y = x^2$

       b)

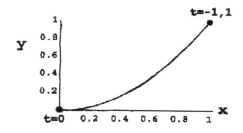

c) No. The curve is not smooth at $t = 0$ where $x'(t)$ and $y'(t)$ are simultaneously zero.

44.   a)

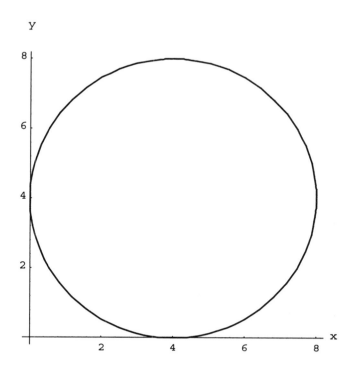

b)

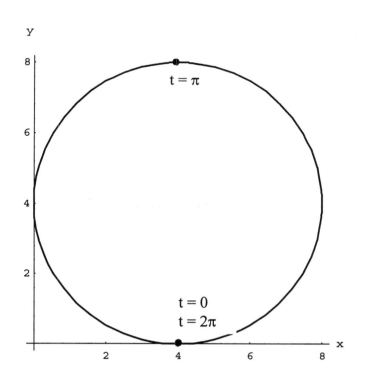

c) −1

45. a) Yes

b) $y = 1 - 2x^2$

46. a)

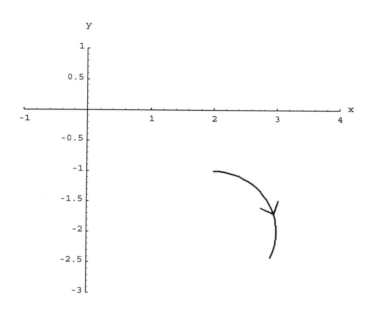

b)

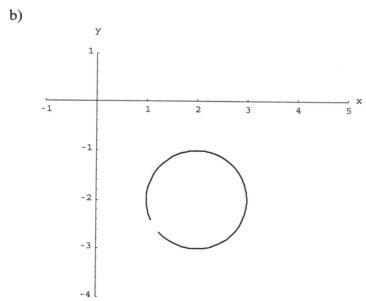

c) No. As t increases, the curve is traced in the same direction. The curve may have different starting and ending points now.

d) Yes. $(x-2)^2 + (y+2)^2 = 1$

47.   a)

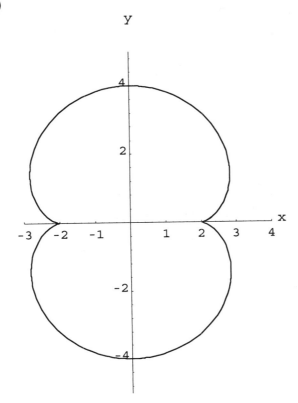

b) $y = \sqrt{3}\,x - 4$

c)   Horizontal tangent at: $(0, 4),\ (0,-4)$

Vertical tangent at: $(2\sqrt{2}, \sqrt{2}),\ (2\sqrt{2}, -\sqrt{2}),\ (-2\sqrt{2}, \sqrt{2}),\ (-2\sqrt{2}, -\sqrt{2})$

SECTION 10: WHY CONTINUITY MATTERS

48.   Can't be true. If the product of the two values is negative they must have opposite signs. One is positive, and one is negative. By the Intermediate Value Theorem, there must be a number c in $(2, 7)$ such that $f(c) = 0$. This contradicts the fact that f has no roots in $[2,7]$.

49.   d) 4   by the Intermediate Value Theorem.

50.   Since $g(-3) = -27.93$ and $g(0) = 1$, there is a point c in $[-3, 0]$ such that $g(c) = -17.1$.

51.   The graph shows that f attains both negative and positive values in [1, 6]. Therefore, by the Intermediate Value Theorem, f must have a root in the interval.

52.   a) Since $f(0) = -1$ and $f(2) = 27$ and f is continuous on [0, 2], there must be a number c in (0, 2) such that $f(c) = 0$

        b) $f(1) = -5$, so a root must be in (1, 2) since 0 is between -5 and 27.

SECTION 11: WHY DIFFERENTIABILITY MATTERS; THE MEAN VALUE THEOREM

53.   Must be true. The Mean Value Theorem guarantees that there is a number c in (3, 7) such that $f'(c) = \dfrac{f(7) - f(3)}{7 - 3} = -\dfrac{3}{4}$. This can be seen in the graph given below. The function $f(x) = -.25x^2 + 1.75x + 2$ satisfies the conditions stated in the problem, and has the appropriate slope at $x = 5$.

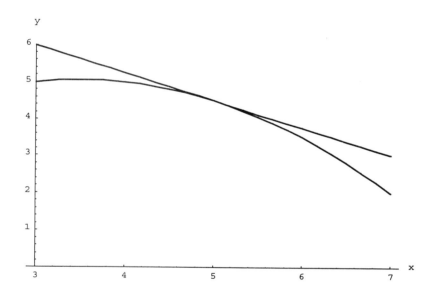

54. $c = \sqrt{\dfrac{7}{3}}$

        A graph is given below.

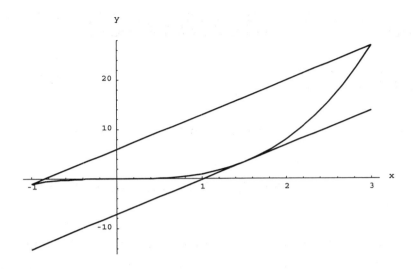

55. Mean Value Theorem: Suppose that f is continuous on the closed interval $[a,b]$ and differentiable on the open interval (a,b). Then for some c between a and b,

$$f'(c) = \frac{f(b) - f(a)}{b - a}.$$

By the Mean Value Theorem, there is a number c in (a, b) such that:

$\dfrac{f(b) - f(a)}{b - a} = f'(c) < 0$, since $f'(x) < 0$ on [a, b]. Therefore,

$f(b) - f(a) < 0$, since $b - a$ is positive. Clearly, $f(b) < f(a)$.

56. $c = \frac{5}{2}$

57. g is **not** differentiable at $x = 0$ in $(-1, 1)$.

143

CHAPTER 5: THE INTEGRAL

SECTION 1: AREAS AND INTEGRALS

1. $\displaystyle\int_{-4}^{5} h = 2\pi - \frac{5}{2} \approx 3.78$

2.   Average value $\approx .42$.

3.

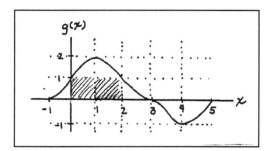

It can be seen from the picture above that the area must be more than 2. The other four parts seem to be about 2, certainly less than 3. Therefore the total must be less than 5.

4. $3 + \frac{\pi}{2} \approx 4.571$

5. -6

6. $\frac{26}{3} + \frac{\pi}{2} \approx 10.237$

7.   a) 7

        b) 17

8.   a) Answers may vary. One possible answer is the piecewise function given by:

$$f(x) = \begin{cases} 11, & 0 \le x < 1 \\ -4, & 1 \le x < 2 \\ 4, & 2 \le x < 4 \end{cases}$$

The graph of f is given below.

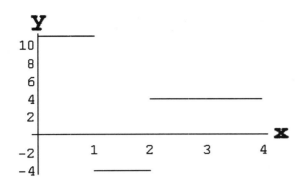

b) 4

c) 11

d) −4

e) If f was always positive or 0 in [1, 2], the area would be positive or 0. This contradicts the given information.

f) If $f(x) < 4$ for all x in [2, 4], then $\int_2^4 f(x)\,dx < \int_2^4 4\,dx = 8$. This contradicts the given information.

SECTION 2: THE AREA FUNCTION

9. The integral will be largest when $c = 3$. After 3, the signed area is negative and will make the integral smaller.

10. a) $A_g(x) = cx$, $A_h(x) = \frac{1}{2}x^2$, $A_i(x) = x^2$, $A_j(x) = \frac{3}{2}x^2$

   b) $\frac{d}{dx}A_g(x) = c$, $\frac{d}{dx}A_h(x) = x$, $\frac{d}{dx}A_i(x) = 2x$, $\frac{d}{dx}A_j(x) = 3x$

   c) The derivative of each area function was the original function.

   d) Conjecture: $\frac{d}{dx}A_f(x) = f(x)$

11. a) $(-\infty, -.312), (1.423, \infty)$

   b) $(-.312, 1.423)$

c) $\left(.556, \infty\right)$

d) $(-\infty, .556)$

e) $-.312,\ 1.423$

12.   a) $\left(-.843, .593\right), \left(4, \infty\right)$

      b) $\left(-\infty, -.843\right), \left(.593, 4\right)$

      c) $\left(-\infty, -.186\right), \left(2.686, \infty\right)$

      d) $\left(-.186, 2.686\right)$

      e) $-.843,\ .593,\ 4$

13. f is dotted, A_f is solid. The graph of f should be 0 when the graph of A_f has a horizontal tangent line. The graph of f should be positive when the graph of A_f is increasing and negative when the graph of A_f is decreasing.

14. f is solid, A_f is dotted. The graph of f should be 0 when the graph of A_f has a horizontal tangent line. The graph of f should be positive when the graph of A_f is increasing and negative when the graph of A_f is decreasing.

SECTION 3: THE FUNDAMENTAL THEOREM OF CALCULUS

15. $k = 3$

16. $\sin(x^6)$

17.   a) 4

      b) -6

c)

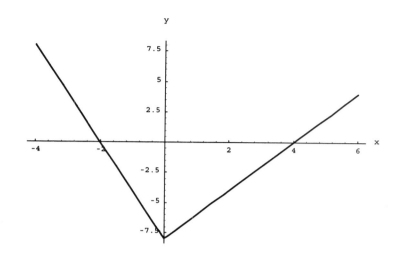

d) $F(x) = \begin{cases} -4x - 8, & x \le 0 \\ 2x - 8, & x > 0 \end{cases}$

e) -4

f) 2

18.　$f(x) = -x^3 + \frac{15}{2}x^2 + \frac{25}{4}$

19.　0

20.　a) $f'(x) = x\cos x + \sin x$

b) $2\pi^2 \approx 19.7392$

SECTION 4: APPROXIMATING SUMS: THE INTEGRAL AS A LIMIT

21.　a) $-x$

b) $-.005$

c) $-.00466343$; error: $.00033657$

22. $L_3 = 3 + \sin 1 + \sin 2 \approx 4.751$; actual: $\frac{11}{2} - \cos 3 \approx 6.49$

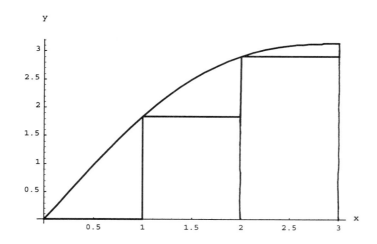

23. $\displaystyle\int_0^8 e^x \, dx$

24.   a) 16

        b)

n, number of subintervals	R_n	absolute error	relative error
1	48	32	2
2	24	8	.5
4	18	2	.125

25.   a) .64

        b) .64

SECTION 5: APPROXIMATING SUMS: INTERPRETATIONS AND APPLICATIONS

26.   11.6 miles

27. a)

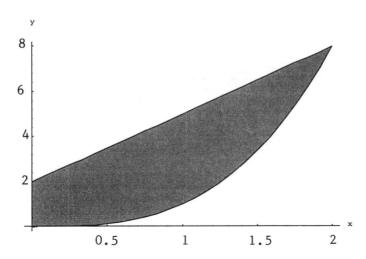

b) 6

c) 6.25

d) 6

28.

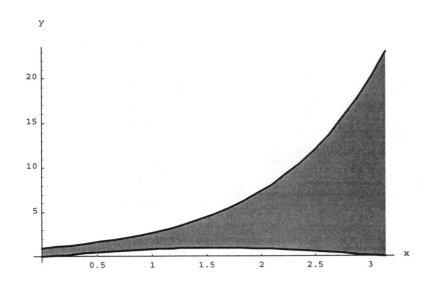

Area $= e^{\pi} - 3 \approx 20.1407$

29.

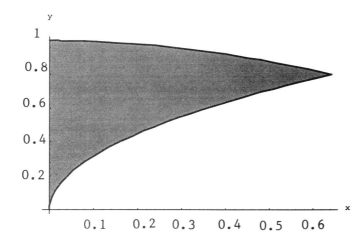

Area ≈ .25586

30.

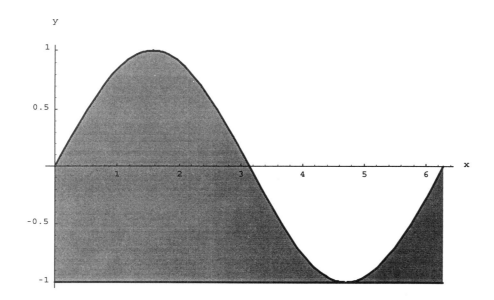

Area = 2π

CHAPTER 6: FINDING ANTIDERIVATIVES

SECTION 1: ANTIDERIVATIVES: THE IDEA

1. $\frac{3}{2}x^2 + \frac{1}{3}e^{3x} + C$

2. $\frac{1}{5}\ln|x| + \dfrac{1}{5x} + \frac{1}{5}x + C$

3. $3\pi \arcsin\left(\frac{x}{\pi}\right) + C$

4. $\dfrac{d}{dx}\left[-\frac{1}{2}\ln\left|\dfrac{2+3x}{x}\right| + C\right] = -\frac{1}{2} \cdot \dfrac{1}{\frac{2+3x}{x}} \cdot \dfrac{x(3)-(2+3x)}{x^2} = -\frac{1}{2} \cdot \dfrac{x}{2+3x} \cdot \dfrac{3x-2-3x}{x^2}$

$$= -\frac{1}{2} \cdot \dfrac{x}{2+3x} \cdot \dfrac{-2}{x^2} = \dfrac{1}{x(2+3x)}$$

5. $\dfrac{d}{dx}\left[2\tan^{-1}\sqrt{2x-1} + C\right] = 2 \cdot \dfrac{1}{1+(2x-1)} \cdot \frac{1}{2}(2x-1)^{-\frac{1}{2}} \cdot 2 = 2 \cdot \dfrac{1}{2x}(2x-1)^{-\frac{1}{2}} = \dfrac{1}{x\sqrt{2x-1}}$

6. Form 1: $\dfrac{d}{dx}\left[-\frac{1}{6}\cos 3x \sin 3x + \frac{1}{2}x\right] = -\frac{1}{6}\left[\cos 3x \cos 3x \cdot 3 - \sin 3x \cdot 3 \cdot \sin 3x\right] + \frac{1}{2}$

$$= -\frac{1}{6}\left[3\left(\cos^2 3x - \sin^2 3x\right)\right] + \frac{1}{2}$$

$$= -\frac{1}{2}\left(1 - \sin^2 3x - \sin^2 3x\right) + \frac{1}{2} = -\frac{1}{2}\left(1 - 2\sin^2 3x\right) + \frac{1}{2}$$

$$= -\frac{1}{2} + \sin^2 3x + \frac{1}{2} = \sin^2 3x$$

Form 2: $\dfrac{d}{dx}\left[\frac{1}{2}x - \frac{1}{12}\sin 6x\right] = \frac{1}{2} - \frac{1}{12}\cos 6x \cdot 6 = \frac{1}{2} - \frac{1}{2}\cos 6x = \frac{1}{2} - \frac{1}{2}\left(1 - 2\sin^2 3x\right)$

$$= \frac{1}{2} - \frac{1}{2} + \sin^2 3x = \sin^2 3x$$

SECTION 2: ANTIDIFFERENTIATION BY SUBSTITUTION

7. $a = 1,\ b = e^{10} \approx 22026.4658,\ \text{Integral} = \frac{1}{5}\left(\sin e^{10} - \sin 1\right) \approx -.30604$

8. $\frac{7}{c}$

9. $\frac{1}{2}\tan e^{2x} + C$

10. $-\arcsin(\cos x) + C$

11. $\dfrac{3\pi}{16} \approx .58905$

12. $\dfrac{1}{\ln 2} \approx 1.4427$

SECTION 3: INTEGRAL AIDS: TABLES AND COMPUTERS

13. $\frac{1}{30}\left(5e^{2x} + 14\right)^3 + C$

14. $\sqrt{e^{2x} - 4} + C$

15. $-\frac{1}{9}\left(3\cos 2x + 1\right)^{\frac{3}{2}} + C$

16. $\frac{1}{2}\cos\left(x^2 - 3x\right) - \frac{1}{6}\cos\left(3x^2 - 9x\right) + C$

17. $\tan x - \frac{1}{2}\ln\left|1 - \pi e^{2\tan x}\right| + C$

CHAPTER 7: NUMERICAL INTEGRATION

SECTION 1: THE IDEA OF APPROXIMATION

1.   a) $\ln 2 \approx .6931472$

       b) $T_{10} \approx .6937714$

       c) $\left| I - T_{10} \right| \approx .0006242$

       d) $\left| I - T_{10} \right| \leq .025$

2.

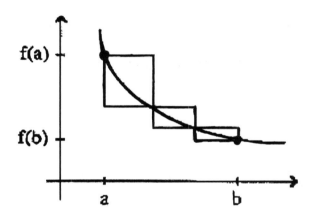

The picture above shows the error from using R_3, the underestimate, and L_3, the overestimate. The difference $\left| R_3 - L_3 \right|$ is just the sum of the areas of the three rectangles. If we stack the rectangles on top of each other, it is easy to compute the area. The width is $\dfrac{b-a}{3}$ and the total height is $\left| f(b) - f(a) \right|$. Thus, $\left| R_3 - L_3 \right| = \left| f(b) - f(a) \right| \cdot \dfrac{(b-a)}{3}$.
Since the function is monotone, the actual value of I must be between R_3 and L_3. So,

$$\left| I - L_3 \right| \leq \left| f(b) - f(a) \right| \cdot \frac{(b-a)}{3} \text{ and } \left| I - R_3 \right| \leq \left| f(b) - f(a) \right| \cdot \frac{(b-a)}{3}.$$

3. $R_3 = \frac{13}{12}$

4.   a) $L_5 = 1$

b)

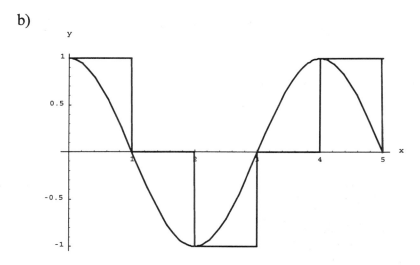

c) $|I - L_5| \le 5$

d) $I = \frac{2}{\pi}$, Actual error $= \left|1 - \frac{2}{\pi}\right| \approx .36338$

5. a) $R_{10} \approx 13.75895$

b) $R_{10} = \frac{1}{2} \sum_{i=1}^{10} \sqrt{\frac{3}{2} i}$

6. a) $L_3 = 1 + e + \dfrac{1}{e} \approx 4.08616,$ $R_3 = 1 + \dfrac{1}{e} + \dfrac{1}{e^2} \approx 1.503215$

b) $T_3 \approx 2.794688$

c) $I = e - \dfrac{1}{e^2} \approx 2.5829465$

d) $|I - L_3| = 1 + \dfrac{1}{e} + \dfrac{1}{e^2} \approx 1.5032147,$ $|I - R_3| = e - 1 - \dfrac{1}{e} - \dfrac{2}{e^2} \approx 1.079732,$
$|I - T_3| \approx .2117415$

e) The error in using T_3 is much smaller than the error in using either L_3 or R_3.

SECTION 2: MORE ON ERROR: LEFT AND RIGHT SUMS AND THE FIRST DERIVATIVE

7. a) $R_4 \approx .555$

   b)

   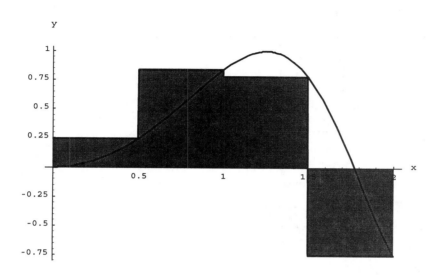

   c) $|I - R_4| \leq 2$ Note: Answers may vary.

8. a) $I = \frac{16}{3}$

   b) $L_{10} = \sqrt{0} \cdot (.4) + \sqrt{.4} \cdot (.4) + \sqrt{.8} \cdot (.4) + \ldots + \sqrt{3.6} \cdot (.4)$

   $R_{10} = \sqrt{.4} \cdot (.4) + \sqrt{.8} \cdot (.4) + \sqrt{1.2} \cdot (.4) + \ldots + \sqrt{4} \cdot (.4)$

   c) $L_{10} = (.4)^{\frac{3}{2}} \sum\limits_{i=0}^{9} \sqrt{i}$

   d) $T_{10} = 5.28407$

   e) increasing

   f) small, concave down

   g) The error bound cannot be calculated since the derivative does not exist at $x = 0$.

   h) $|I - L_{10}| = \frac{16}{3} - 4.88407 \approx .44926$

9. a) $L_{20} = \frac{3}{20} \sum_{i=0}^{19} (.5)^{\frac{3i}{20}}$

    b) large, decreasing

    c) Theorem 1: $|I - L_{20}| \leq .13125$

        Theorem 2: $|I - L_{20}| \leq .155958$

10. a) $L_6 \approx 1.6993798$

    b) $|I - L_6| \leq .643323$

11. a) E_2 is larger since e^x is steeper.

    b) $|I_1 - L_2| = \frac{5}{3}$, $|I_2 - L_2| = e^2 - e - 2 \approx 2.670774$ which is larger

12. a) $K_1 = 2.5$, so $|I - R_{10}| \leq 12.5$

    b) No! The function f is not monotone since f' is both positive and negative on [0, 10].

SECTION 3: TRAPEZOID SUMS, MIDPOINT SUMS, AND THE SECOND DERIVATIVE

13.

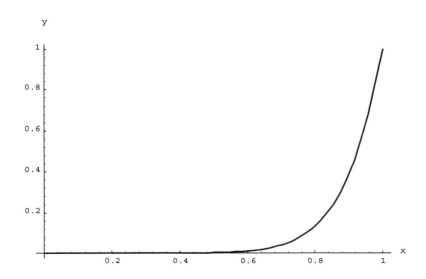

a) $I = .1$

b) $L_1 = 0, \quad T_1 = .5$

c)   Error using $L_1 = .1$, error using $T_1 = .4$ which is larger. Although the trapezoid method is a better approximation for large n, this doesn't contradict anything.

d) $L_2 \approx .0009765625$, Error $\approx .099023$

   $T_2 \approx .25098$, Error $\approx .15098$, which is larger

   $L_3 \approx .0086877$, Error $\approx .0913123$

   $T_3 \approx .17535$, Error $\approx .07535$ which is smaller.

   The trapezoid approximation is better when $n = 3$.

14.   a)   .5

   b) $T_5 \approx .5057831$

   c)   Error $\approx .0057831$

   d) $\left| I - T_5 \right| \leq .075$

   e) $\left| I - T_5 \right| \leq .02$

15. $L_n = 9.61105, \quad M_n = 9.73601, \quad T_n = 9.73677, \quad R_n = 9.86249$

   The left sum is the smallest of the four and the right sum is the largest of the four since f is increasing. The midpoint sum is an underestimate, and the trapezoid sum is an overestimate, since f is concave up.

16. $T_4 = 29.5$ miles

17.   a) $M_2 = \frac{16}{15} \approx 1.06667$

   b) $T_2 = \frac{7}{6} \approx 1.16667$

   c)   Yes, since f is monotone.

d) $K_1 = 1$

e) Using Theorem 1: $|I - L_{100}| \leq \frac{2}{150} \approx .01333$

Using Theorem 2: $|I - L_{100}| \leq \frac{1}{50} = .02$

18. $T_4 \approx 1.38652$

SECTION 4: SIMPSON'S RULE

19. a) $I = .5(1 - \cos 4) \approx .826821810432$

b) $S_{20} \approx .826829195104$

c) Error $= .0000073846722$

d) $|I - S_{20}| \leq .0000177777$. As expected, the actual error is less than this theoretical bound.

e) $n = 6$

20. a) $L_4 = 1.4675$, $M_2 = 1.495$, $T_4 = 1.4625$, $S_4 = 1.47333$

b) $1.4625 \leq I \leq 1.495$, since the trapezoid sum underestimates and the midpoint sum overestimates functions which are concave down.

21. a) $M_2 \approx 10.77176$, $T_2 \approx 30.517356845$, $S_4 \approx 17.3536264504$

b) $|I - S_4| \leq 15.621$

c) $n = 142$

22. a) 557. If $K_1 = 3.6$ is used, the answer is $n = 566$

b) 14. If $K_2 = 9$ is used, the answer is $n = 15$

c) 20. If $K_2 = 9$ is used, the answer is $n = 21$

d) 7

23. a) $S_2 = \frac{4}{3} \approx 1.333$

b) $K_4 = 24$

c) $|I - S_2| \leq \frac{4}{15} \approx .26667$

d) Actual error $= \frac{4}{15}$

e) $n = 15$

24. a) $R_4 = \frac{3\pi}{4} \approx 2.35619449$

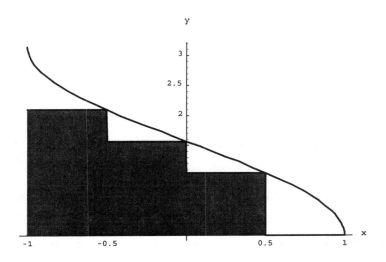

b) small, since f is decreasing.

c) $T_4 = \pi$, $M_4 = \pi$

d) $S_4 = \pi$

e) The derivative does not exist at 1 or -1. K_1 is undefined.

f) Yes, they are exact! The overestimate of one rectangle or trapezoid is always canceled out by an underestimate in another rectangle or trapezoid.

159

CHAPTER 8: USING THE DEFINITE INTEGRAL

SECTION 1: INTRODUCTION

1.   a) Curve: $\sqrt{1+\cos^2 x}$

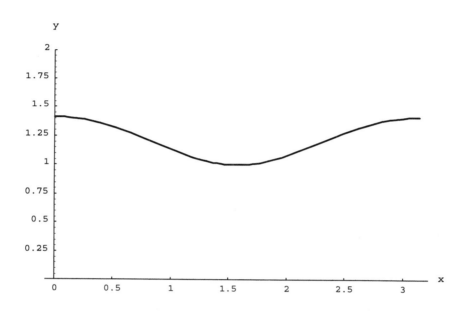

Using a crude estimation, we could approximate the area under the curve as the area of a rectangle with width 3 and height 1.25. This yields an approximate area of 3.75. Note: answers will vary depending on the approximation method used.

b) Function: $f(x) = \sin x$

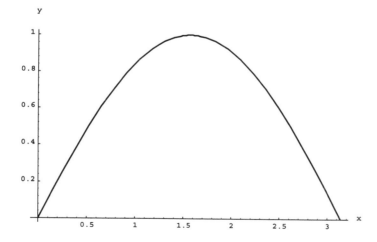

Using a crude approximation, we could approximate the arc length as the sum of the lengths of two line segments. One segment goes through (0, 0) and $\left(\frac{\pi}{2},1\right)$. The other segment goes through $\left(\frac{\pi}{2},1\right)$ and $(\pi,0)$. This method yields an approximate arc length of $2\sqrt{1+\dfrac{\pi^2}{4}} \approx 3.72419$.

c) $T_{100} \approx 3.82019778903$

2. a) $y = \left(\dfrac{y_2 - y_1}{x_2 - x_1}\right)(x - x_1) + y_1$

b) $f'(x) = \dfrac{y_2 - y_1}{x_2 - x_1}$

c) $\sqrt{1 + \left(\dfrac{y_2 - y_1}{x_2 - x_1}\right)^2} \cdot (x_2 - x_1)$

d) $\sqrt{1 + \left(\dfrac{y_2 - y_1}{x_2 - x_1}\right)^2} \cdot (x_2 - x_1) = \sqrt{\dfrac{(x_2 - x_1)^2 + (y_2 - y_1)^2}{(x_2 - x_1)^2}} \cdot (x_2 - x_1)$

$$= \sqrt{(x_2 - x_1)^2 + (y_2 - y_1)^2}$$

which is the result given by the distance formula.

SECTION 2: FINDING VOLUMES BY INTEGRATION

3. a)

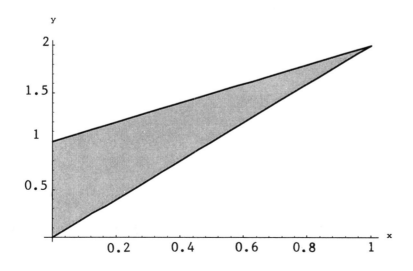

b) Volume = π

4.   Volume = $\frac{16}{3}$

5.   Volume = $\frac{5\pi}{24}$

6.   a) Volume = $\int_f^e \left[\pi \cdot l^2(y) - \pi \cdot h^2(y)\right] dy$

       b) Volume = $\int_f^e \left[\pi \cdot \left(l(y) - g\right)^2 - \pi \cdot \left(h(y) - g\right)^2\right] dy$

7.   Volume = $288\sqrt{3}$

8.   Volume = $\frac{8\pi}{3}$

SECTION 3: ARCLENGTH

9. $\int_{.5}^3 \sqrt{1 + \frac{1}{x^2}}\, dx$

10.   Length = $\frac{422}{5}$

11.   Arc length = $2\sqrt{3}$

12.   a)

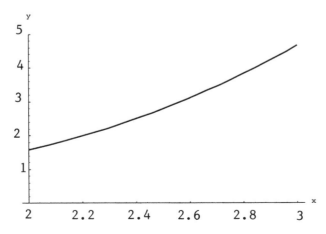

       Approximating the curve by a straight line with endpoints (2, 1.6) and (3, 4.7) yields an arc length of 3.26.

b) Arc length = 3.25

13. Arc length = $e^2 - 1 \approx 6.389$

SECTION 4: WORK

14. Work = .42 ft-lb.

15. a) Spring constant = 12 lb. / ft

b) Work = 54 ft-lb.

c) Work = 1200 ft-lb.

16. Work = 31,250 ft-lb.

17. Work = $62.4 \int\limits_{0}^{20} 64\pi(7+x)\,dx$

18. Length = 1.5 ft

19. Work = 17664π ft-lb.

20. Work = 276 ft-lb.

SECTION 5: PRESENT VALUE

21. $r = \frac{\ln 3}{10} \approx .10986 \approx .11$

22. PV = $36,744.32

23. $p(t) = 100\sqrt{t}$ is the rate of income flow at time t, in dollars per year. The 2 and the 4 tell us to find the present value between the second and fourth years. $r = .07$ means an annual interest rate of 7%

24. .2086, about 21% !

CHAPTER 9: MORE ANTIDIFFERENTIATION TECHNIQUES

SECTION 1: INTEGRATION BY PARTS

1. $\frac{1}{2}x^2 \ln 5x - \frac{1}{4}x^2 + C$

2. $x \ln 5x - x + C$

3. $\dfrac{x^{100}}{100} \cdot \ln x - \dfrac{x^{100}}{10000} + C$

4. $e^x \left(x^3 - 3x^2 + 6x - 6\right) + C$

5. $\frac{1}{3}x^3 \ln \sqrt{x} - \frac{1}{18}x^3 + C$

6. $\frac{1}{26}e^t \left(\sin 5t - 5\cos 5t\right) + C$

7. $-\frac{1}{x}\ln x - \frac{1}{x} + C$

8. $-x\cos x + \sin x + C$

SECTION 2: PARTIAL FRACTIONS

9. $\frac{1}{2}\ln|x-1| + \frac{5}{2}\tan^{-1}x - \frac{1}{4}\ln(1+x^2) + C$

10. $\frac{6}{5}\ln|x-2| - \frac{12}{5}\tan^{-1}x - \frac{3}{5}\ln(1+x^2) + C$

11. $-\dfrac{1}{1+x} + 2\ln|1+x| + C$

12. $\ln 2$

13. $\ln|x-3| - \ln|x-2| + C$

14. $\frac{1}{5}\ln|x| - \frac{1}{10}\ln(5+x^2) + C$

15. $-2\ln|2+x| + 3\ln|3+x| + C$

16. $-\int \dfrac{1}{x-1} dx + 2\int \dfrac{1}{x} dx + 3\int \dfrac{1}{2+x} dx$

SECTION 3: TRIGONOMETRIC ANTIDERIVATIVES

17. $\frac{1}{6}\tan^{-1}\left(3x^2\right) + C$

18. $\frac{1}{2}\ln\left(x^2 + \sqrt{x^4 - 25}\right) + C$

19. $\frac{\pi}{12}$

20. $\frac{1}{2}\ln\left(x^2 + 2\right) + \sqrt{2}\tan^{-1}\left(\dfrac{x}{\sqrt{2}}\right) + C$

21. $\frac{1}{2}x\sqrt{25 - x^2} + \frac{25}{2}\sin^{-1}\left(\frac{x}{5}\right) + C$

22. $-\frac{1}{2}\cos(2x) + \frac{1}{6}\cos^3(2x) + C$

23. $\left(-6 - \dfrac{x^2}{3}\right)\sqrt{9 - x^2} + C$

CHAPTER 10: IMPROPER INTEGRALS

SECTION 1: WHEN IS AN INTEGRAL IMPROPER?

1.   Diverges

2.   Converges. Integral $= \ln(1 + \sqrt{2})$

3.   a) The integral is improper because the interval is infinite.

     b) Diverges

4.   a) The integral is improper because the interval is infinite.

     b) Converges. Integral $= 1$

5.   a) The integral is improper because the integrand is unbounded at 0.

     b) Diverges

6.   Diverges

7.   Converges. Integral $= \frac{4}{3}$

8.   Choosing $u = \dfrac{1}{x}$ means $du = -\dfrac{1}{x^2}dx$. Choosing $dv = xe^{-x^2}dx$ means $v = -\frac{1}{2}e^{-x^2}$.
Now, by parts:

$$\int_a^\infty e^{-x^2}dx = -\frac{1}{2x}e^{-x^2}\Bigg]_a^\infty - \frac{1}{2}\int_a^\infty \frac{1}{x^2}e^{-x^2}dx = \frac{e^{-a^2}}{2a} - \frac{1}{2}\int_a^\infty \frac{e^{-x^2}}{x^2}dx$$

SECTION 2: DETECTING CONVERGENCE, ESTIMATING LIMITS

9. $\displaystyle\int_1^\infty \frac{1}{\sqrt{1+x^3}}\,dx \le \int_1^\infty \frac{1}{\sqrt{x^3}}\,dx = 2$. So the integral converges.

10. $\displaystyle\int_2^\infty \frac{x}{\sqrt{x^3-2}}\,dx \ge \int_2^\infty \frac{x}{\sqrt{x^3}}\,dx$ which diverges. So the original integral also diverges.

11. $\displaystyle\int_0^\infty \frac{e^{-x}}{\sqrt{x}}\,dx = \int_0^1 \frac{dx}{e^x\sqrt{x}} + \int_1^\infty \frac{dx}{e^x\sqrt{x}} \le \int_0^1 \frac{dx}{\sqrt{x}} + \int_1^\infty \frac{dx}{e^x} = 2 + \frac{1}{e}$. So, the original integral converges.

12. $\displaystyle\int_1^\infty \frac{4}{x(x+1)}\,dx \le \int_1^\infty \frac{4}{x^2}\,dx = 4$. So, the original integral converges.

13. $\displaystyle\int_0^\infty \frac{dx}{x^3 + e^{-x}}\,dx = \int_0^1 \frac{dx}{x^3 + e^{-x}}\,dx + \int_1^\infty \frac{dx}{x^3 + e^{-x}}\,dx \le \int_0^1 \frac{dx}{e^{-x}} + \int_1^\infty \frac{dx}{x^3} = e - \frac{1}{2}$. So, the original integral converges.

SECTION 3: IMPROPER INTEGRALS AND PROBABILITY

14. a) $\displaystyle\frac{1}{5\sqrt{2\pi}} \int_{62}^{70} e^{-\frac{(x-75)^2}{50}}\,dx$

b) $\displaystyle\int_2^{2.4} \frac{1}{\sqrt{2\pi}}\, e^{-\frac{x^2}{2}}\,dx$

15. a) $e^{-.4} - e^{-.8} \approx .22$

b) 2.5 minutes

c) $1 - e^{-.4x}$

16. a) The probability that a person will live for at most t years after the treatment.

b) $S(t) = \displaystyle\int_t^\infty p(x)\,dx = e^{-Ct}$

c) $C = -\dfrac{\ln .7}{2} \approx .178337$

SECTION 4: L'HOPITAL'S RULE: COMPARING RATES

17. Converges. Integral $= 1$.

18. The limit does not exist.

19. 0

20. $\frac{1}{2}$

21. $\frac{2}{3}$

22. ln 3

23. 0

24. 0

CHAPTER 11: INFINITE SERIES

SECTION 1: SEQUENCES AND THEIR LIMITS

1.   The sequence converges since the limit exists. The sequence converges to $\frac{3}{2}$.

2.   The sequence converges since the limit exists. The sequence converges to $\sqrt{7}$.

3.   a) $a_n = -10^{3-n}$

        b) $a_{101} = -\dfrac{1}{10^{98}}$

        c)   The sequence converges since $\lim\limits_{n \to \infty} a_n = 0$

        d)   0

4.   a)   all reals

        b)   The sequence converges to $-k$

5.   a)   all reals except 0

        b)   The sequence converges to $-\frac{1}{k}$

6.   a)   all reals

        b)   The sequence converges to -1

SECTION 2: INFINITE SERIES, CONVERGENCE, AND DIVERGENCE

7.   The series diverges since $a_n \to \frac{3}{2}$ as $n \to \infty$ (the n-th term test).

8.   The series diverges since the terms do not tend to 0 as $n \to \infty$ (the n-th term test).

9.   The series converges to $\sin 1$ by the n-th sum test. $S_n = \sin 1 - \sin \frac{1}{n+1} \to \sin 1$ as $n \to \infty$.

10.   The series converges to 3. This can be shown using the geometric series test (since $|r| < 1$).

11. a) $S_{10} = 7.5$

    b) 7

    c) 0, since the series converges.

12. a) might be true: Ex. 1: $\{n\}$ is increasing and $\sum n$ diverges

                             Ex. 2: $\left\{-\dfrac{1}{n^2}\right\}$ is increasing and $\sum\left(-\dfrac{1}{n^2}\right)$ converges

    b) might be true: Ex. 1: $\left\{\dfrac{1}{n^2}\right\}$ is decreasing and $\sum\left(\dfrac{1}{n^2}\right)$ converges

                             Ex. 2: $\{\frac{1}{n}\}$ is decreasing and $\sum\frac{1}{n}$ diverges

    c) can't be true: If $\{a_n\}$ is increasing and positive then $a_n \geq a_1 > 0$ $\forall n$.
                             So, the terms can't tend to 0 as $n \to \infty$. By the n-th term test, the
                             series must diverge.

    d) might be true: Ex. 1: $\dfrac{1}{n^2} \to 0$ and $\sum\left(\dfrac{1}{n^2}\right)$ converges

                             Ex. 2: $\frac{1}{n} \to 0$ and $\sum\frac{1}{n}$ diverges

13. The series converges to $-\frac{1}{4}$. This can be shown using the n-th sum test.

$$S_n = \frac{1}{(n+2)^2} - \tfrac{1}{4} \to -\tfrac{1}{4} \text{ as } n \to \infty.$$

14. $\dfrac{2\pi^2 + 7}{12}$

15. Consider a right approximating sum with n subdivisions. Then $\Delta x = \frac{1}{n}$ and the height of the kth rectangle is determined by the right endpoint of the subinterval: $\frac{k}{n}$. Adding the areas of the rectangles, we see that $R_n = \sum\limits_{k=1}^{n} \frac{k}{n} \cdot \frac{1}{n} = a_n$. Thus, $\lim\limits_{n\to\infty} a_n = \int\limits_0^1 x\,dx$.

SECTION 3: TESTING FOR CONVERGENCE; ESTIMATING LIMITS

16. The series converges by the ratio test.

17. The series diverges. $\sum \dfrac{k!7^k}{2^k} \geq \sum \dfrac{7^k}{2^k} = \sum \left(\dfrac{7}{2}\right)^k$ which is a divergent geometric series since $|r| > 1$.

18. The series converges. $\sum \dfrac{\sqrt{k}}{k^3+1} < \sum \dfrac{\sqrt{k}}{k^3} = \sum \dfrac{1}{k^{\frac{5}{2}}}$ which is a convergent p-series.

19. The series diverges. $\sum \dfrac{\ln k}{k} > \sum \dfrac{1}{k}$ since $\dfrac{\ln k}{k} > \dfrac{1}{k}$ for $k > 3$.

20. The series diverges. $\displaystyle\sum_{k=1}^{\infty} \dfrac{\ln k}{k} = 0 + \dfrac{\ln 2}{2} + \sum_{k=3}^{\infty} \dfrac{\ln k}{k}$, so we only need to test the last term.

    But, $\displaystyle\sum_{k=3}^{\infty} \dfrac{\ln k}{k} \geq \int_3^{\infty} \dfrac{\ln x}{x}\,dx = \lim_{t \to \infty} \int_3^t \dfrac{\ln x}{x}\,dx = \lim_{t \to \infty} \tfrac{1}{2}(\ln x)^2 \Big|_3^t$ which does not exist.

21. The ratio test is inconclusive since $\dfrac{\ln(k+1)}{k+1} \cdot \dfrac{k}{\ln k} \to 1$

22. No, since $\dfrac{k^2}{(k+1)^2} \to 1$

SECTION 4: ABSOLUTE CONVERGENCE; ALTERNATING SERIES

23. Using the alternating series test, we see that 3 terms of the series will be sufficient to approximate the sum to within .005. The approximation is $S_3 \approx -.2579$.

24. a) might be true   Ex. 1: $\sum \left(\dfrac{1}{n^2}\right)$ converges and $\sum \left|\dfrac{1}{n^2}\right|$ converges

    Ex. 2: $\sum \left(-\tfrac{1}{k}\right)$ converges and $\sum \left|-\tfrac{1}{k}\right| = \sum \tfrac{1}{k}$ diverges

    b) must be true   The alternating signs make it easier to converge.

25. a) The series converges.

    b) The comparison test and the integral test.

26. a) The series diverges.

    b) The n-th term test.

27.   a) The series converges.

        b) alternating series test

        c) conditional by the p-series test

28.   a) The series converges.

        b) geometric series test or alternating series test

        c) absolute by the geometric series test

29.   a) The series converges.

        b) alternating series test

        c) conditional by the comparison test

30.   a) The series converges.

        b) alternating series test

        c) conditional by integral test

SECTION 5: POWER SERIES

31. $[-3, 1)$

32. $[-1, 5)$

33. $[-1, 1]$

34. $\{0\}$

35. $[-1, 1]$

36. $\{4\}$

37.   a) might be true

        b) might be true

        c) must be true

d) may be true

38. $\left(-\frac{5}{4}, \frac{5}{4}\right)$

SECTION 6: POWER SERIES AS FUNCTIONS

39. $x - \dfrac{x^2}{2^2} + \dfrac{x^3}{3^2} - \dfrac{x^4}{4^2} + \dots$

40. $x + x^2 + x^3 + x^4 + x^5 + \dots$ on $(-1, 1)$

41. $x^2 - \dfrac{x^4}{2!} + \dfrac{x^6}{4!} - \dfrac{x^8}{6!} + \dots$ on $(-\infty, \infty)$

42. $x + \dfrac{x^2}{2!} - \dfrac{x^3}{3!} - \dfrac{x^4}{4!} + \dfrac{x^5}{5!} + \dfrac{x^6}{6!} - \dots$ on $(-\infty, \infty)$

43. $\dfrac{x^4}{4} - \dfrac{x^6}{6 \cdot 3!} + \dfrac{x^8}{8 \cdot 5!} - \dfrac{x^{10}}{10 \cdot 7!} + \dots$ on $(-\infty, \infty)$

44. e^{x^2}

45. $y = e^{x^2} = 1 + x^2 + \dfrac{x^4}{2!} + \dfrac{x^6}{3!} + \dots$

So, $y' = 2x + 4\dfrac{x^3}{2!} + 6\dfrac{x^5}{3!} + 8\dfrac{x^7}{4!} + \dots = 2x + 2x^3 + x^5 + 2\dfrac{x^7}{3!} + \dots = 2xy$.

SECTION 7: MACLAURIN AND TAYLOR SERIES

46. a) $(x-1)^2 - (x-1)^3$

b) $\frac{5}{192} \approx .02604$.

47. a) $\displaystyle\sum_{k=0}^{\infty} (-x)^k = 1 - x + x^2 - x^3 + x^4 - \dots$

b) $x^2 - x^5 + x^8 - x^{11} + x^{14} - \dots$

c) $x^3 - \dfrac{x^6}{2} + \dfrac{x^9}{3} - \dfrac{x^{12}}{4} + \ldots$

48.   a) $\displaystyle\sum_{k=0}^{\infty} \dfrac{(3x)^k}{k!}$

    b) $(-\infty, \infty)$

49. $(-1,1)$. A graph of f and S_{20} is shown below:

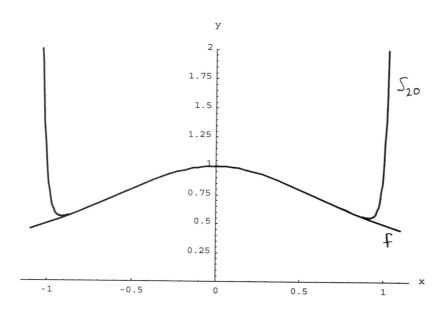

50. $(0, 2]$. A graph of f and S_{10} is shown below:

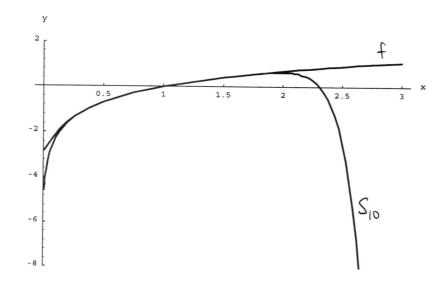

174

51. a) $x - \dfrac{x^2}{2 \cdot 3!} + \dfrac{x^3}{3 \cdot 5!} - \dfrac{x^4}{4 \cdot 7!} + \dfrac{x^5}{5 \cdot 9!} - \cdots$

b) Making the substitution $u = \sqrt{x}$, we see that $du = \dfrac{1}{2\sqrt{x}} dx$ so:

$$\int \frac{\sin \sqrt{x}}{\sqrt{x}} dx = 2 \int \frac{\sin \sqrt{x}}{2\sqrt{x}} dx = 2 \int \sin u \, du = -2 \cos u = -2 \cos \sqrt{x}$$

The Maclaurin series for $-2 \cos \sqrt{x}$ is $-2 + x - \dfrac{x^2}{2 \cdot 3!} + \dfrac{x^3}{3 \cdot 5!} - \dfrac{x^4}{4 \cdot 7!} + \dfrac{x^5}{5 \cdot 9!} - \cdots$

The two series differ by a constant, which is acceptable for two antiderivatives.

CHAPTER 12: DIFFERENTIAL EQUATIONS

SECTION 1: DIFFERENTIAL EQUATIONS: THE BASICS

1.   Yes. It is a solution, since y is continuous and differentiable and it satisfies the DE.

2.   No. The left hand side of the DE is $y' = \frac{1}{2}e^x - 2e^{-2x}$, which is not equal to the right hand side: $2y + e^{-2x} = e^x + 3e^{-2x}$.

3.   Yes. It is a solution, since y is continuous and differentiable and it satisfies the DE.

4.   a) $y(t) = 5e^{9t}$

       b) $\dfrac{dy}{dt} = 5e^{9t} \cdot 9 = 9y$

5.   a) $y(x) = x^2 + 3$

       b) $y(x) = 3e^{2x}$

SECTION 2: SLOPE FIELDS: SOLVING DE'S GRAPHICALLY

6. $y(x) = x^3 - x^2 - x + 10$

7.   a)

b)

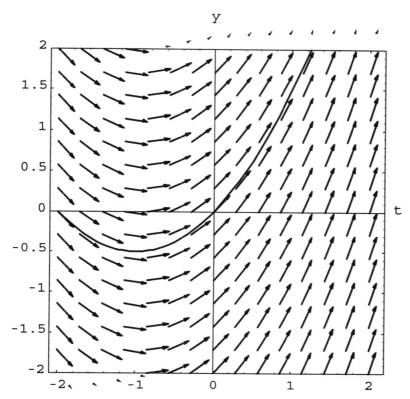

c) $y(t) = \frac{1}{2}t^2 + t$, which agrees with the picture

8.   a) iii

      b) iv

      c) i

      d) vi

      e) ii

      f) v

9.   a) $y = 1$, $C = 0$ (see the picture in part (d))

      b) $C = -3$, since $y(0) = 1 - C = 4$. (see picture in part (d))

c) The tangent line seems to go through $(-2, 10)$ and $(0, 4)$, so slope is -3. This is what the DE predicts.

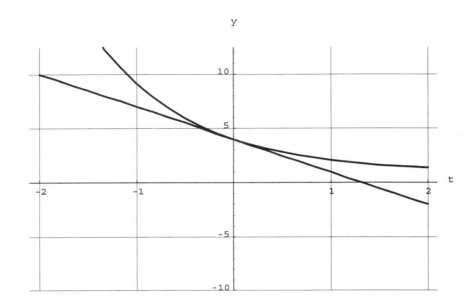

d)

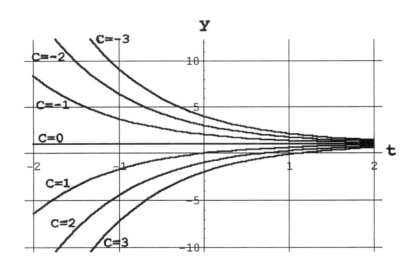

SECTION 3: EULER'S METHOD: SOLVING DE'S NUMERICALLY

10. $Y(0) = 2$, $Y(.5) = 3$, $Y(1) = 4.25$, $Y(1.5) = 5.875$, $Y(2) = 8.0625$

11. a) $y(t) = e^{-2t}$, so $y(1) = e^{-2} \approx .135335$, $y(2) = e^{-4} \approx .018316$, $y(3) = e^{-6} \approx .002479$, and $y(4) = e^{-8} \approx .000335$

b) $Y(1) = -1$, $Y(2) = 1$, $Y(3) = -1$, $Y(4) = 1$. The estimates are not very good.

12. a) $y(t) \equiv 0$. $y(1) = y(2) = y(3) = y(4) = 0$

b) $Y(1) = Y(2) = Y(3) = Y(4) = 0$. Euler's method predicts the zero solution, which agrees with (a).

SECTION 4: SEPARATING VARIABLES: SOLVING DE'S SYMBOLICALLY

13. $y(t) = 500 + C e^{-.8t}$

14. a) If P represents the population: $P' = kP(1{,}000 - P)$.

b) $P(t) = \dfrac{1{,}000}{19e^{-.3736072t} + 1}$

15. a) Left hand side of DE: $P' = -C\left(1 + 2e^{-3Ct}\right)^{-2}\left(-6Ce^{-3Ct}\right) = \dfrac{6C^2 e^{-3Ct}}{\left(1 + 2e^{-3Ct}\right)^2}$

Right hand side of DE: $3P(C - P) = \dfrac{3C}{1 + 2e^{-3Ct}}\left(C - \dfrac{C}{1 + 2e^{-3Ct}}\right)$

$$= \dfrac{3C}{1 + 2e^{-3Ct}}\left(\dfrac{C + 2Ce^{-3Ct} - C}{1 + 2e^{-3Ct}}\right)$$

$$= \dfrac{6C^2 e^{-3Ct}}{\left(1 + 2e^{-3Ct}\right)^2}$$

b) C

c) $P(0) = \dfrac{C}{3}$, $P(1) = \dfrac{C}{1 + 2e^{-3C}}$

d) $P'' = 3P'(C - 2P)$. Setting $P'' = 0$, we find that $P = \frac{C}{2}$ is an inflection point.

e) $t = \dfrac{\ln .5}{-3C}$

16. $y(t) = e^{.5t^2}$

17. a) $y(t) = 50 + 50e^{-.102165t}$

    b) $t \approx 15.7533$

18. a) Left hand side of DE: $\dfrac{dy}{dt} = \sec t \tan t$

    Right hand side of DE: $y^2 \sin t = \sec^2 t \sin t = \dfrac{\sin t}{\cos^2 t} = \sec t \tan t$

    b) $y(t) = \dfrac{1}{\cos t + 1}$

19. a) $\dfrac{dT}{dt} = k(T - 10)$

    b) $T(t) = 10 + 58e^{-.023366t}$. The temperature will be approximately 48°F. If this model is correct, you should not worry about the pipes freezing.

    c) The assumption is made that the outside temperature remains constant. The actual temperature in the house is probably lower since the temperature outside will probably drop during the night.

20. a) $P(t) = 800 - 300e^{kt}$

    b) $k = .5 \ln \frac{1}{3} \approx -.549306$

    c) 800

CHAPTER 13: POLAR COORDINATES

SECTION 1: POLAR COORDINATES AND POLAR CURVES

1. $r = \sec\theta \tan\theta$

2. $(2\sqrt{3}, 2)$

3. $\left(2, -\frac{\pi}{6} + 2n\pi\right), \quad n = 0, \pm 1, \pm 2, \pm 3, \ldots$

4. $r = 2a\cos\theta$

5.   The curve is a parabola. In rectangular coordinates: $y = \frac{1}{4}x^2 - 1$.

6.   The intersection points are: the origin, $\left(\frac{1}{2}, \frac{\pi}{3} + 2n\pi\right)$ and $\left(\frac{1}{2}, \frac{5\pi}{3} + 2n\pi\right)$, for $n = 0, \pm 1, \pm 2, \ldots$

   In rectangular coordinates, the intersection points are: $(0,0)$, $\left(\frac{1}{4}, \frac{\sqrt{3}}{4}\right)$, $\left(\frac{1}{4}, -\frac{\sqrt{3}}{4}\right)$.

7.   The curve is shown below. The only line of symmetry is the x-axis.

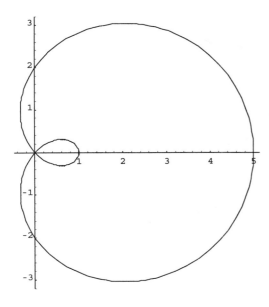

8.   The curve is shown below:

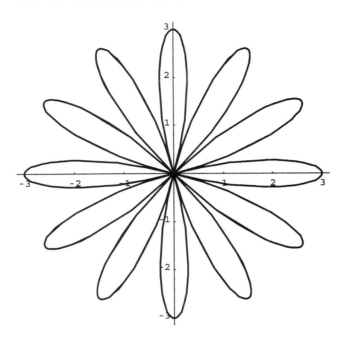

The lines of symmetry are the x-axis, the y-axis, $y = \pm\frac{\sqrt{3}}{3}x$ and $y = \pm\sqrt{3}\,x$, which are shown in the graph below:

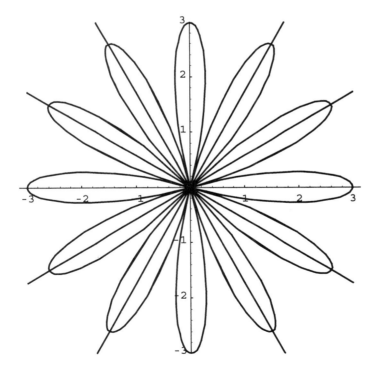

9.   The curve is shown below. Significant points are: $(-6,0)$, $(-2,0)$, $(0, 0)$, $(0, 2)$, and $(0,-2)$.

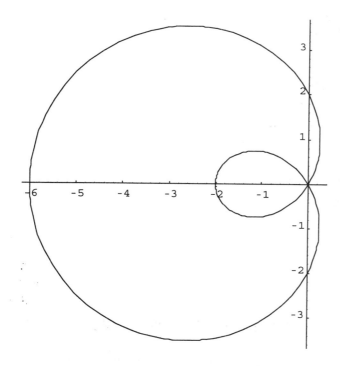

10.   The curve is shown below. Significant points are: $(-2,0)$, $(2, 0)$, $(0, 0)$, and $(0,-4)$.

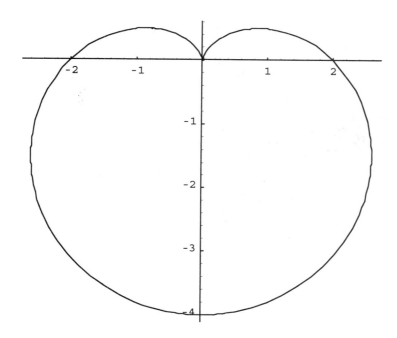

11.   The curve is shown below. Note: This is the graph of the vertical line $x = -5$.
        Significant points are: $(-5, 0)$.

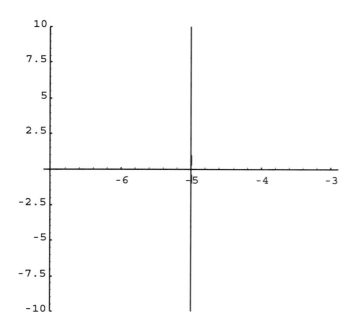

12.   The curve is shown below. Significant points are: $(-1.5, -1.5)$, $(1.5, -1.5)$, $(0, 0)$, and
        $(0, -3)$.

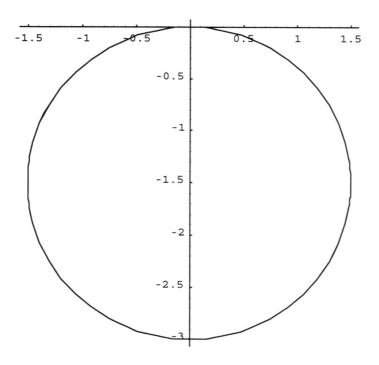

SECTION 2: CALCULUS IN POLAR COORDINATES

13.   Area $= 2\sqrt{3} + \dfrac{4\pi}{3} \approx 7.65289$. The region is shown shaded in the graph below:

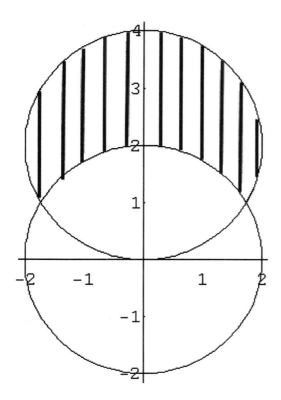

14.   Area $= \frac{3\pi}{2} \approx 4.71239$. The curve is shown below:

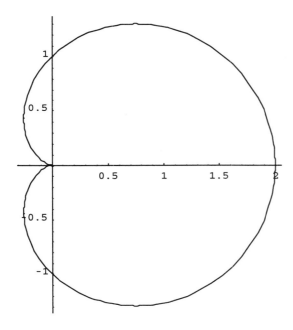

15. a)

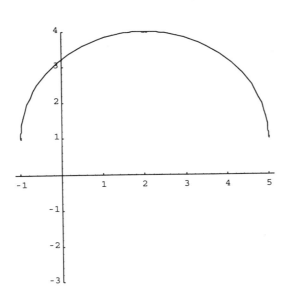

b) $(x-2)^2 + (y-1)^2 = 9$

c) 0

16. a)

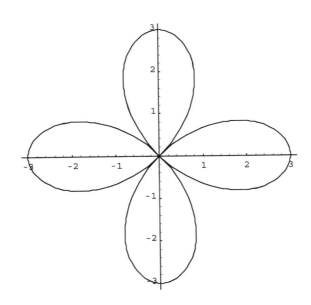

b) $\displaystyle 2\int_0^{\pi/4} \frac{(3\cos(2\theta))^2}{2}\,d\theta = \int_0^{\pi/4} 9\cos^2(2\theta)\,d\theta$

17.   Area $= 4\pi + 12\sqrt{3} \approx 33.35098$. The curve is shown below:

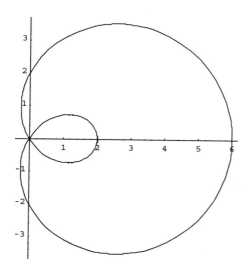

CHAPTER 14: MULTIVARIABLE CALCULUS: A FIRST LOOK

SECTION 1: THREE-DIMENSIONAL SPACE

1.   The graph is shown below:

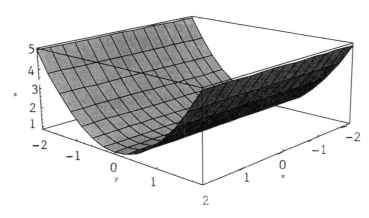

2.   You end at the point $(1, -1, 1)$.

   a) in front of

   b) to the left of

   c) above

3.   a) $y^2 + z^2 = 4$

   b)

4. a)

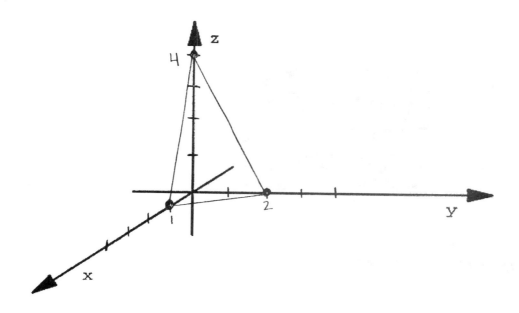

b) $z = -4x + 4$

5. a) $\sqrt{106}$

b) $(x+1)^2 + (y-7)^2 + z^2 = 106$

6. $3x - 2y = 5$

7.

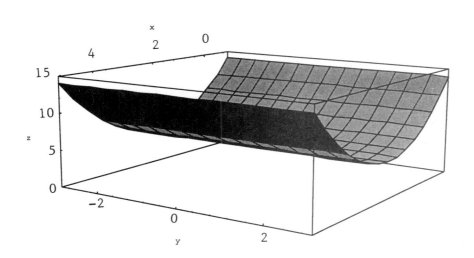

8.   Note: The cross sections are circles with center $(-4, 3)$ and radius $2\sqrt{10}$.

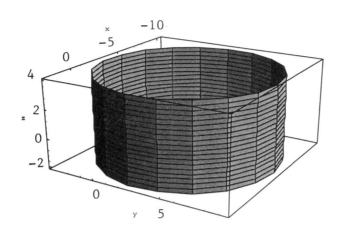

SECTION 2: FUNCTIONS OF SEVERAL VARIABLES

9.

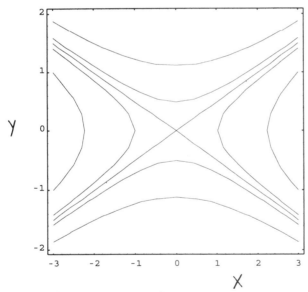

10.   a) $\left\{ (x, y) \mid |x| \geq |y| \right\}$

       b) $\left\{ z \mid z \geq 0 \right\}$

11.   a) $\Re^3$

       b) $\left\{ w \mid w \geq 0 \right\}$

12. a) $\left\{ (x, y) \mid (x, y) \neq (0,0) \right\}$

b) $\left\{ z \mid z > 0 \right\}$

c)

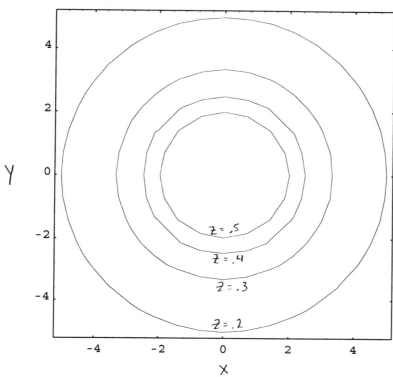

13. a)

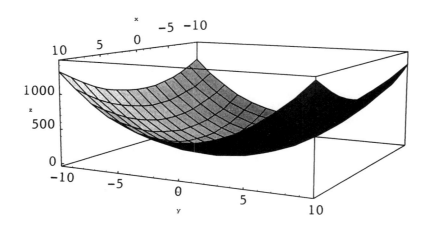

b) Level curves are ellipses.

c) $\dfrac{(x+1)^2}{\frac{3}{2}} + \dfrac{\left(y+\frac{1}{3}\right)^2}{\frac{2}{3}} = 1$

SECTION 3: PARTIAL DERIVATIVES

14. $z = 2x - 8y - 9$

15.   a) $f_w(w,15)$ for $5 < w < 10$

b) $f_w(10,25) \approx -1.6$ This means that when the temperature is $25°$ and the wind speed is 10 mph, the wind chill factor drops by about $1.6°$ for each 1 mph increase in wind speed.

c) $f_T(20,5) \approx -1.6$ and $f_w(20,5) \approx -1$

d) $L(w,T) = -3 - w - \frac{8}{5}T$

16.   a) $h_x = 8y - (x+y)^3$

b) $h_y = 8x - (x+y)^3$

c) The graph of f is shown below:

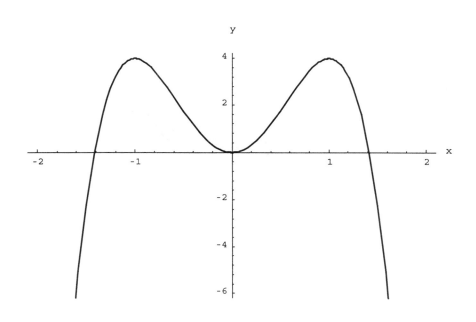

The graph of g is shown below:

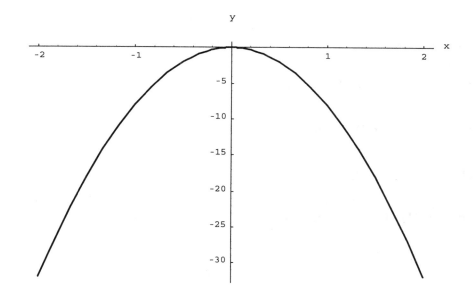

17. a) 6

b) 4

c) 1

d) 1

e) 2

f) 5

g) 6

h) 5

i) 4

j) 4

k) 1

l) 2

Function 3 is not represented.

18.   a) $f_x(1,-1)=2, \quad f_y(1,-1)=2$

        b) $f_x(1,-1)=-2, \quad f_y(1,-1)=3$

        c) $f_x(1,-1)=-\frac{2}{9}, \quad f_y(1,-1)=\frac{2}{9}$

        d) $f_x(1,-1)=-2, \quad f_y(1,-1)=-3$

        e) $f_x(1,-1)=\frac{1}{\sqrt{2}}, \quad f_y(1,-1)=-\frac{1}{\sqrt{2}}$

        Note: In parts (f) - (i), answers may vary.

        f) $f_x(1,-1)=.83, \quad f_y(1,-1)=-.41$

        g) $f_x(1,-1)=-2, \quad f_y(1,-1)=3.5$

        h) $f_x(1,-1)\approx.6$

        i) $L(x,y)=.666-.166x+.167y$

19.   a)

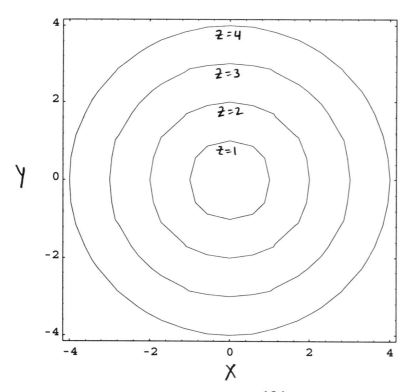

b) Note: Answers may vary. $f_x(1,0) \approx \frac{2-1}{1} = 1$, $f_y(1,1) \approx \frac{2.25-1.25}{1} = 1$

c) $f_x(x,y) = \dfrac{x}{\sqrt{x^2 + y^2}}$, $f_y(x,y) = \dfrac{y}{\sqrt{x^2 + y^2}}$

d) $f_x(1,0) = 1$, $f_x(1,1) = \frac{1}{\sqrt{2}}$

e) $L(x,y) = \frac{1}{\sqrt{2}}x + \frac{1}{\sqrt{2}}y$.   The linear approximation is the tangent plane to $f(x)$ at $(1, 1)$.

SECTION 4: OPTIMIZATION AND PARTIAL DERIVATIVES: A FIRST LOOK

20. f attains a minimum of $-\sqrt{5}$ at $\left(-\frac{1}{\sqrt{5}}, -\frac{2}{\sqrt{5}}\right)$.

21.   The maximum value is 1, which is attained at $(0, 0, 1)$.

22.   The stationary points are: $(.5, 1)$, $(-.5, -1)$, which are saddles, $(.5, -1)$, which is a local minimum, and $(-.5, 1)$, which is a local maximum.

23.   a)   The z values are given in the graph below.

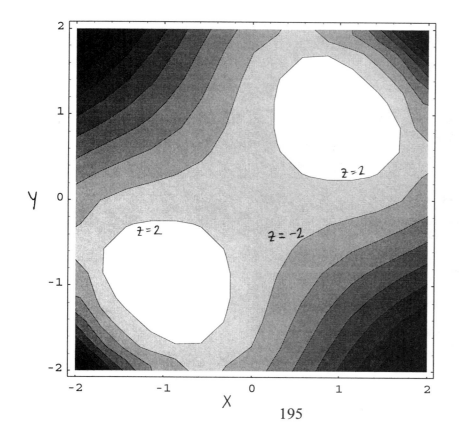

b) The stationary points are: $(0, 0)$ is a saddle, $(1, 1)$ is a local maximum, and $(-1, -1)$ is a local maximum.

24.   a) $g_x(x, y) = (2x - 2)\sin y$

b) $g_y(x, y) = (x^2 - 2x)\cos y$

c) The dark dot in the light part of the contour plot is a local maximum and the light dot in the dark part of the contour plot is a local minimum.

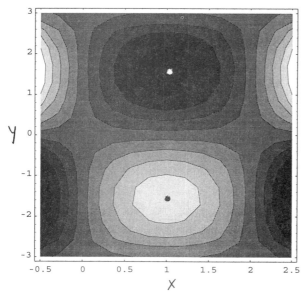

d) Both are saddles.

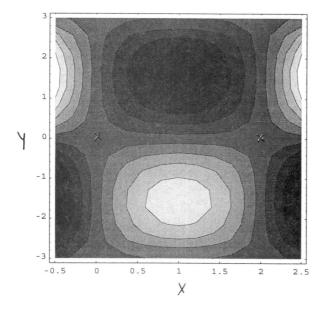

e) The stationary points are: $\left(1, \frac{\pi}{2}\right)$, $\left(1, -\frac{\pi}{2}\right)$, $(0,0)$, $(2,0)$

25. The magnetic field has a minumum of 12 at $(-4, 4, 2)$ and $(-4, -4, 2)$

SECTION 5: MULTIPLE INTEGRALS AND APPROXIMATING SUMS

26. 13.5   Note: Answers may vary.

27. a) The volume under the curve is approximately 24. This figure was obtained by considering just one quarter of the surface at a time. In the 2 by 2 by 3 rectangular solid (volume = 16) defined by x in [0, 2], y in [0, 2], and z in [0, 3], the graph seems to cut the total volume approximately in half. So, the volume under the curve, in this quarter, is approximately 6. The same argument holds for the other 3 quarters, thus the total estimate for volume is 24.

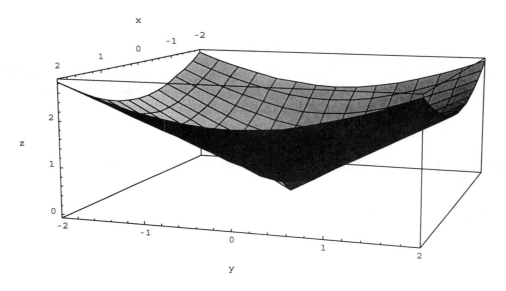

b) $6\sqrt{2} + 4\sqrt{10} + 4\sqrt{\frac{1}{2}} \approx 23.96$

28. a) The volume under the curve is approximately 8. In the 2 by 2 by 4 rectangular solid (volume = 16) defined by x in [0, 2], y in [0, 2], and z in [0, 4], the graph seems to cut the total volume approximately in half. So, the volume under the curve is approximately 8.

197

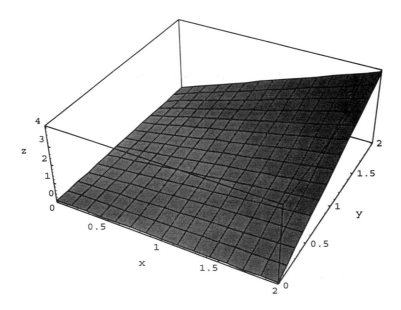

b) 1

c) In each of the four subdivisions, the rectangular solid using the point nearest the origin for height underapproximates the actual volume. Therefore, the estimate given in (b) is smaller than the actual volume.

SECTION 6: CALCULATING INTEGRALS BY ITERATION

29. $V = \int_0^3 \int_0^{2x} x\, y^2\, dy\, dx = 129.6$

30. a)

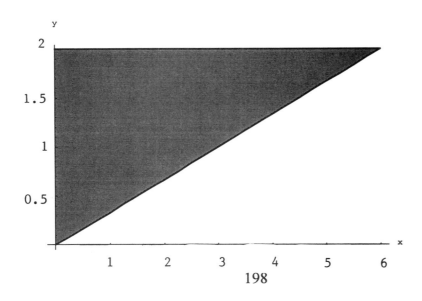

b) 26

31. $\frac{1}{3}(1 - \cos 8) \approx .381833$

SECTION 7: DOUBLE INTEGRALS IN POLAR COORDINATES

32. $\frac{\pi}{4}(1 - \cos 9) \approx 1.501$

33. 2π

34. Area $= 1 + \frac{\pi}{8} \approx 1.3927$. The region is shown below:

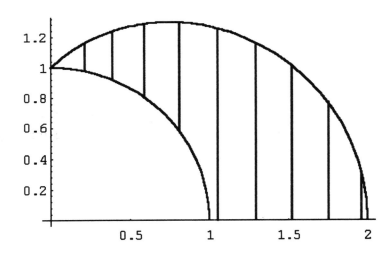